Domain-Specific Bodies of Knowledge in
Project Management – **Volume 1**

# Building a Body of Knowledge in Construction Project Delivery, Procurement and Contracting

**Domain-Specific Bodies of Knowledge in Project Management**

Print ISSN: 2811-0889
Online ISSN: 2811-0897

Series Editor: Mohan M Kumaraswamy
*(The University of Hong Kong, Hong Kong)*

---

*Published:*

Vol. 1 *Building a Body of Knowledge in Construction Project Delivery, Procurement and Contracting*
edited by Giovanni C Migliaccio (University of Washington, USA) and Pramen P Shrestha (University of Nevada Las Vegas, USA)

Domain-Specific Bodies of Knowledge in
Project Management – **Volume 1**

# Building a Body of Knowledge in Construction Project Delivery, Procurement and Contracting

**Editor-in-Chief:** Mohan M Kumaraswamy
The University of Hong Kong, Hong Kong

Editors

Giovanni C. Migliaccio
University of Washington, USA

Pramen P. Shrestha
University of Nevada Las Vegas, USA

World Scientific

NEW JERSEY • LONDON • SINGAPORE • BEIJING • SHANGHAI • HONG KONG • TAIPEI • CHENNAI • TOKYO

*Published by*

World Scientific Publishing Co. Pte. Ltd.
5 Toh Tuck Link, Singapore 596224
*USA office:* 27 Warren Street, Suite 401-402, Hackensack, NJ 07601
*UK office:* 57 Shelton Street, Covent Garden, London WC2H 9HE

**Library of Congress Cataloging-in-Publication Data**
Names: Migliaccio, Giovanni C. (Giovanni Ciro), 1968– editor. |
Shrestha, Pramen P., 1964– editor.
Title: Building a body of knowledge in construction project delivery, procurement and contracting / editors, Giovanni C. Migliaccio, University of Washington, USA, Pramen P. Shrestha, University of Nevada Las Vegas, USA.
Description: New Jersey : World Scientific, 2023. | Series: Domain-specific bodies of knowledge in project management ; volume 1 | Includes bibliographical references.
Identifiers: LCCN 2022031930 | ISBN 9789811224775 (hardcover) |
ISBN 9789811224782 (ebook for institutions) | ISBN 9789811225710 (ebook for individuals)
Subjects: LCSH: Building--Superintendence.
Classification: LCC TH438 .B7955 2023 | DDC 690.068--dc23/eng/20220823
LC record available at https://lccn.loc.gov/2022031930

**British Library Cataloguing-in-Publication Data**
A catalogue record for this book is available from the British Library.



For any available supplementary material, please visit
https://www.worldscientific.com/worldscibooks/10.1142/11952#t=suppl

Desk Editor: Nicole Ong

Typeset by Stallion Press
Email: enquiries@stallionpress.com

# About the Editors

**Pramen P. Shrestha** is a Professor of Construction Engineering Management and Graduate Coordinator in the Civil & Environmental Engineering and Construction department at the University of Nevada Las Vegas (UNLV). He holds an M.Sc. in Civil Engineering from Oklahoma State University and a Ph.D. in Civil Engineering from the University of Texas at Austin, United States (US). He completed his B.Sc. in Civil Engineering from the National Institute of Technology, Jamshedpur, India. He had about 10 years of professional experience in design and construction of highway and infrastructure projects in Nepal before moving to the US. After completing his Ph.D. in Civil Engineering from the University of Texas at Austin in 2007, he joined UNLV as an Assistant Professor. He has received federal, state, and private research grants worth about US $2.5 million. His research grants are funded by the US State Departments of Transportation, Transportation Research Board, National Science Foundation, US Labor Department, Water Design Build Council (WDBC), Airport Cooperative Research Program (ACRP) and private agencies. He has helped with the graduation of five Ph.D. and 34 master's students as an advisor. He currently has two Ph.D. students and three master's students working for him on various projects. He has published more than 55 peer-reviewed journal papers and 75 conference proceedings. He is a Fellow of American Society of Civil Engineers and a licensed professional engineer in Texas.

**Giovanni C. Migliaccio** holds a Howard S. Wright Endowed Professorship, he is the current Chair of the Department of Construction Management, and he has served as the Executive Director of the Center for Education and Research in Construction (CERC). He holds an M.S. and a Ph.D. in Civil Engineering from the University of Texas at Austin and a master-level degree from Politecnico di Bari in Italy. Prior to moving to the US, he worked in Italy in the construction management of telecommunication projects under Nortel Networks, Nokia Networks, and IPSE 2000. At UW, Giovanni is active in all three pillars of academic life, including teaching, research, and service to the university and the industry.

# Preface

In-depth knowledge on project delivery, procurement, and contracting is critical for any project manager to effectively perform during the pre-planning, design, construction, operation, and maintenance phases of any type of facilities. This book intends to consolidate, structure, and articulate collective knowledge on construction project delivery, procurement, and contracting, to initiate building a body of knowledge in this domain.

Early on, to successfully complete any building or infrastructure project, the owner's project manager needs to select an appropriate delivery method, which "defines the relationships, roles, and responsibilities of project team members and the sequence of activities required to complete a project" (Gibson and Walewski, 2001, p. 1). This selection is usually based on the characteristics and risks inherent to the project and the desired owner's control over design and construction, while in public works, it is also more constrained by what is allowed by relevant legislation and regulations.

The chosen delivery method influences the procurement methods used to purchase external services and materials necessary to deliver the project. This involves identifying and selecting designers or builders, so that the project can be successfully designed, built, operated, and maintained. During the selection of the designer or builder, it is paramount to select an appropriate contracting process for establishing a contractual relationship for services and/or materials through the development of a written agreement, expressing the expectations, rights, responsibilities, and protections of each party. The selection of an adequate contracting process will enhance the chance of completing the project within budget, schedule, and with quality. Therefore, individuals who aspire to become successful project managers need to have in-depth knowledge in these areas which are covered in this book.

As researchers in construction project delivery, procurement, and contracting, we often need to educate our doctoral and master students on this broad area of construction engineering and management. This means relying on miscellaneous sources for their initial education. Similarly, we often interact with professionals who have a partial understanding of the underlying principles and ever-changing best practices worldwide. Therefore, we need to provide them with a broad overview before supporting their efforts. The main intent of this book is to consolidate, structure, and articulate a core 'body of knowledge' on this critical theme, so that it can serve as a foundation to education and as a reference book for professionals. We all know that many specific and valuable reference books exist on sub-topics, such as design–build, diverse contracting approaches, and procurement selection, but an umbrella contribution that can act as gateway introduction to an integrated body of knowledge on construction project delivery, procurement, and contracting is lacking. In addition to supporting the education of our own student researchers and broadening and deepening the knowledge of practitioners, this book is also intended to be used as a textbook by us, our colleagues, and many of the programs that incorporate an undergraduate or graduate course on the topics of project delivery, procurement, and contracting.

This book includes fifteen chapters organized into three sections. The initial section covers the fundamentals of project delivery and is followed by sections on procurement and contracting approaches. Each section will first provide a general overview of the concepts that will be later discussed in detail in various chapters in that section.

Section A consists of five chapters, namely project delivery methods, integrated digital delivery, project environment, early contractor involvement in construction projects, and an owner's perspective on delivering green buildings. Section B starts with a summary of procurement approaches used during design, construction, and material procurement. Then, the qualifications-based selection process and best-value procurement process will be discussed followed by procurement strategy for

# Contents

# Part A

# Project Delivery

# Chapter 1

# Project Delivery Methods

Giovanni C. Migliaccio[1] and Pramen P. Shrestha[2]

[1]*University of Washington*
[2]*University of Nevada Las Vegas*

## Background

All construction projects share similarities, as they usually evolve along a comparable lifecycle and rely on a set of participants that share similar roles. Still, the project owner's approach to delivery affects participants' interactions, lifecycle phase overlaps, and even project financing. Throughout the rest of this book, we will refer to project delivery method (PDM) as the owner's selected approach for delivering a project, affecting what services need to be procured from outside entities, which will become project participants, and the contracting terms of these relationships.

In evaluating their needs to identify project objectives, owners also evaluate what services they can self-perform and what they will need to contract out. While increasing the number of contractual relationships may help owners to secure the best providers for various project components, this also increases their responsibilities associated with coordinating everyone's effort, so they all stay aligned to the same project objectives. To avoid the need for coordination among external entities, project owners may be tempted not to have any contracts and self-perform all of the work, relying on their own staff to design and construct a project; however, this delivery scenario is unrealistic for most construction projects, and its adoption is limited to specific types of projects and owners. Instead, contracting with external entities is usually the main characteristic of construction project delivery.

When owners contract out for project services, they are usually procuring external expertise on the project; a written contractual agreement is adopted to convey the owner's expectations and the rights and responsibilities of the contractual parties. Typical contracting approaches for built environment services will be discussed in Section C. Contracting out may occur in different ways and at different project phases, with the selected PDMs identifying when each contract would be in place throughout the project's lifecycle. When multiple contracts are concurrently executed, the need for coordination among contracted providers increases.

The selection of a PDM is usually made early on, as it significantly affects the project by shaping the roles and responsibilities among the project's team members; further, it affects the lifecycle itself, along with the number of services to be contracted out as well as the number of external entities to be brought onto the project and when their selection occurs. Project delivery is becoming increasingly important, as it also intertwines with how owners finance construction projects. The rest of this chapter introduces the main concepts related to project delivery and describes the most commonly used project delivery methods.

## Project phases

Generally, project owners identify their needs and then conclude that a construction project should result in the completion of a facility or infrastructure and its placement into operations in a specific time frame. However, most built environment projects are time consuming and require significant investments, which made them unviable for single phase delivery; thus, phased delivery is quite common in construction, with expenditure authorizations often tied to a phase-gate process to authorize project expenditures.

The major phases in delivering a facility or infrastructure include planning, design, construction, operations, and maintenance; they are usually sequential with overlaps. During the planning phase, project owners realize their needs, evaluate alternative ways to achieve these needs, and generate a project concept. The design phase is usually more resource-intensive and often broken down into sub-phases, such as conceptual design, schematic design, detailed design, and construction document preparation. Construction professionals are usually procured to lead a project's construction phase. During this phase, these professionals first plan and prepare for the construction process, then perform or oversee field activities, and are responsible for the outcome. Once construction is completed, the facility or infrastructure is ready to be put into operation, which generally means that the project owner starts using it.

## Definition of project delivery method

As previously stated, the PDM is the owner's selected approach for delivering a project, affecting what services need to be procured from outside entities, which will become project participants, and the contracting terms of these relationships. A construction project is defined by its objectives and usually includes, at minimum, the completion of its design and physical construction, but could also include objectives beyond the design and construction phases. A short but comprehensive definition of PDM was formulated by Gibson and Walewski (2001) and has been slightly revised as follows: "...a project delivery method [...] defines the relationships, roles, and responsibilities of project team members and the sequence of activities required to complete a project" (p. 1) and achieves its objectives.

While some project delivery methods expand their reach into the operations and maintenance phases, most built environment projects are initiated and planned up to the initial portion of their life span, which ends at the completion of the construction phase; however, models for project delivery that extend their reach into the operations of a facility or infrastructure exist.

## Main models for project delivery

A frequent approach to visualizing PDMs is relying on their contractual structure. The selected PDM will influence all the project team members and how they will work together; the life span will also be affected by this upfront decision. For instance, if an owner decides to have a firm deliver both design and construction services, this firm will enter into an eventually phased contract with the owner to deliver their scope of work. Understanding the features of project delivery and their chemistry provides owners with new ways to deliver and finance construction projects, along with opportunities to introduce and experiment with new practices, and eventually, modify them to their own needs. Here, we limit our discussion to the most common delivery methods even though construction projects are often delivered through new or hybrid approaches.

## *Design–Bid–Build (DBB)*

This method relies on a sequential delivery of design and construction services. Design is initially performed by in-house owner staff or a design firm and results in design documents and specifications that will be used as the basis for soliciting prices for a construction contract. During the bid phase, construction contractors review and price the project based on its design documents and specifications as well as other factors. Those contractors who submit a bid and meet the procurement process requirements are deemed responsive, and the procurement process will determine which contractors are offered a contract. This model of delivery is sequential, with the contractor being selected once the design is nearly complete.

## *Construction Management-at-Risk (CMR)*

When an owner selects CMR, design and construction services are contracted out separately as in DBB; however, construction services are secured during or at the start of design, so that the contractor can review and comment on the design while it is being prepared. The contractor's role is often designed around two steps: (1) as the design is prepared, the contractor often acts as a consultant for pre-construction services, including design review; (2) once the design progresses toward completion, the contractor negotiates a contract for construction services, including a price and project delivery date.

## *Design–Build (DB)*

The popularity of this model has grown significantly mostly due to its ability to integrate delivery of design and construction services under a single contract with an entity, the design–builder, which will design and build the project. For large projects, design–builder entities are frequently joint ventures among several firms that dissolve once the project is completed. The DB model has several variations depending on the selected features for procuring and contracting. Most similar to DBB, *Bridging DB* sees the owner advancing design to create a set of bridging documents before selecting a design–builder; these documents are used as the basis of the contract. Relying on a procurement that heavily weighs price, among other factors, and on a lump sum contract, *price-competitive DB* awards a design–build contract after a portion of the design — usually between 10% and 30% — is completed by the owner. Inspired by CMR, *progressive DB* (PDB) utilizes a two-step approach, so that the owner initially selects an entity to act as the design–builder based on their qualifications and enters into a design and preconstruction services contract with them; once the design progresses toward completion, the design–builder negotiates a contract for construction services, including a price and project delivery date. In PDB, if the owner is unable to negotiate the price of construction, then the owner has a right to terminate the contract with the design–builder and deliver the project using the DBB method.

## *Integrated Project Delivery (IPD)*

IPD is the most recent approach to project delivery. The main difference between IPD and other project delivery methods is its reliance on a multiparty contractual agreement. As defined by the American Institute of Architects (AIA), IPD is "… *a project delivery approach that integrates people, systems, business structures and practices into a process that collaboratively harnesses the talents and insights of all participants to optimize project results, increase value to the owner, reduce waste, and maximize efficiency through all phases of design, fabrication, and construction*" (AIA, 2007, back cover).

## Public–private partnership

It is important to know the owner's strategy for financing a project. Even though financing is not a part of PDMs, based on the financing method, the selection of a PDM can be affected. In the context of a PDM, a project's financing method, as well a conceptual knowledge of the features of the approach for financing a project, is relevant. A project finance method is a system for acquiring or providing funds from different sources and combining them to fund a project. Project finance methods can be differentiated by the following:

- The project's life span, which identifies the part of the life of the corresponding facility that is financed.
- The types of financing sources that provide funding to the project (e.g., owner, lenders, contractor, etc.).
- The types of financing vehicles that are used (e.g., equity, loan, etc.).

Public–private partnership is defined as a method of combining various phases of projects, e.g., finance, planning, designing, construction, operation, and maintenance. While the previously described main models for project delivery only deliver design and construction, PPP models can include additional project phases, including its financing.

## Public statutory framework regarding construction contracts under various delivery methods

Public entities have historically been limited to using DBB because the Brooks Act, enacted in 1972 (Public Law 92-582), limited the procurement of design services on federally funded projects on the basis of qualifications only and the procurement of construction services to the lowest responsible responsive bidder. In general, when owners decided to use an alternative PDM, they needed to make a case and demonstrate the benefits in using these PDMs to the state or federal government.

As per 49 CFR §18.36: Procurement, "[w]hen procuring property and services under a grant, a State will follow the same policies and procedures it uses for procurements from its non-federal funds" (Office of the

Secretary of Transportation, 18.36, p. 127). Thus, in general, if a specific state/city uses an alternative project delivery method, federal regulations do not hinder them from using it. With these many changes over the past few decades, publicly owned projects are increasingly being able to choose alternative PDMs for capital improvement projects.

## Summary

In any type of project, the PDM is the owner's plan of how to deliver their project. Still, a good understanding of the adopted PDM provides a navigation compass to project managers. Today, certain industry sectors and their owners are quickly evolving and pushing the boundaries of project delivery. While in the 1950s, the majority of project management professionals thought they would work for most, if not all, of their careers on DBB projects, today we should adapt to an evolving approach to delivery.

This chapter was a brief introduction to the topic of project delivery, and hopefully, sparked interest in the readers. The forthcoming chapters will continue to build upon these concepts by elaborating more on the issues that are critical in understanding the various project delivery methods. This section of the book covers five chapters related to project delivery. Chapter 2 discusses the obstacles and solutions in adopting an integrated digital delivery defined lifecycle approach that digitally integrates the entire construction value chain by implementing digital technologies. Chapter 3 describes the relationship between project delivery method and project environment. Chapter 4 discusses the drivers and impediments of involving the contractor early on in the delivery of a project. The final Chapter 5, reports what project delivery methods are preferred by green building project owners in the Australian construction industry and provides an insight about how project owners can contribute to the selection and enhancement of project performance.

# About the Authors

**Pramen P. Shrestha** is a Professor of Construction Engineering Management and Graduate Coordinator in the Civil & Environmental Engineering and Construction department at the University of Nevada Las Vegas (UNLV). He holds an M.Sc. in Civil Engineering from Oklahoma State University and a Ph.D. in Civil Engineering from the University of Texas at Austin, United States (US). He completed his B.Sc. in Civil Engineering from the National Institute of Technology, Jamshedpur, India. He had about 10 years of professional experience in design and construction of highway and infrastructure projects in Nepal before moving to the US. After completing his Ph.D. in Civil Engineering from the University of Texas at Austin in 2007, he joined UNLV as an Assistant Professor. He has received federal, state, and private research grants worth about US $2.5 million. His research grants are funded by the US State Departments of Transportation, Transportation Research Board, National Science Foundation, US Labor Department, Water Design Build Council (WDBC), Airport Cooperative Research Program (ACRP) and private agencies. He has helped with the graduation of five Ph.D. and 34 master's students as an advisor. He currently has two Ph.D. students and three master's students working for him on various projects. He has published more than 55 peer-reviewed journal papers and 75 conference proceedings. He is a Fellow of American Society of Civil Engineers and a licensed professional engineer in Texas.

**Giovanni C. Migliaccio** holds a Howard S. Wright Endowed Professorship, he is the current Chair of the Department of Construction Management, and he has served as the Executive Director of the Center for Education and Research in Construction (CERC). He holds an M.S. and a Ph.D. in Civil Engineering from the University of Texas at Austin and a master-level degree from Politecnico di Bari in Italy. Prior to moving to the US, he worked in Italy in the construction management of telecommunication projects under Nortel Networks, Nokia Networks, and IPSE 2000. At UW, Giovanni is active in all three pillars of academic life, including teaching, research, and service to the university and the industry.

# Chapter 2

# Obstacles and Solutions for Adoption of Integrated Digital Delivery

Jasmine Ngo, Bon-Gang Hwang and Wan Yi, Priscilla, Her

*Department of Building, National University of Singapore, Singapore*

## Abstract

Integrated Digital Delivery is a lifecycle approach that digitally integrates the entire construction value chain by implementing digital technologies. It has the potential to improve sustainability performances of construction projects and building operations through lean processes and reduction in wastes. However, the construction industry is known to be resistant toward technology adoption and the implementation rate of Integrated Digital Delivery remains at a nascent stage. Hence, to improve the sustainability performances of construction projects and building operations through Integrated Digital Delivery, this chapter will do the following: (i) introduce the concept and technologies of Integrated Digital Delivery; (ii) investigate the common obstacles faced by practitioners in adopting Integrated Digital Delivery; and (iii) propose feasible solutions to reap the benefits by increasing the implementation rate of Integrated Digital Delivery. The findings elaborated in this chapter are expected to provide a better understanding of the Integrated Digital Delivery approach, the overarching obstacles faced by industry practitioners, and the responding solutions so as to drive the adoption of integrated and collaborative project delivery approaches to facilitate the digital transformation of the construction industry.

## Introduction

Integrated and collaborative project delivery approaches have been developed to integrate the fragmented construction value chain and improve project performances (Hwang *et al.*, 2020). For instance, Virtual Design and Construction (VDC) has been introduced to integrate multidisciplinary models of construction projects (Kunz and Fischer, 2012), while Integrated Project Delivery (IPD) establishes a multi-party agreement for risk and reward sharing to optimize project performance (American Institute of Architects, 2007). These integrated and collaborative project delivery approaches were introduced with the overarching aim of reducing the inefficiencies in work processes and information loss arising from the fragmentation of the value chain, and the lack of complete, accurate, and timely information exchange, which are common issues faced in the global construction industry (Building and Construction Authority and Infocomm Media Development Authority, 2018; Dainty *et al.*, 2001; Froese, 2010; Leviäkangas *et al.*, 2017). Beyond project completion, effective and sustainable building operations require accurate building information from the construction phases (Arayici *et al.*, 2012).

On top of the above challenges, Singapore faces additional challenges of lack of local labor supply and poor industry image (Ng, 2017). Hence, the Singapore government introduced the Construction Industry Transformation Map (CITM) that aims to digitally transform the construction industry by driving the adoption of digital technologies across the building lifecycle (Building and Construction Authority, 2018). One key pillar of the CITM is Integrated Digital Delivery (IDD), which is an integrated and digitalized collaborative project delivery approach that complements the use of digital technologies to streamline work processes and connect stakeholders to deliver an optimized and sustainable outcome for end users (Building and Construction Authority, 2018).

The integration of the entire value chain through the use of emerging technologies such as cyber-physical systems and Building Information Modeling (BIM) can improve the performance of the building and construction industries. Specifically, integrating the value chain improves information transfer, enabling the stakeholders to utilize the information for accurate decision-making throughout the building lifecycle. At the same time, unnecessary delays are also minimized, with improvements in on-site safety, sustainability performances of projects and buildings, and increase in productivity (Building and Construction Authority, 2018; Rashid, 2018). Furthermore, the digitalization and automation of work processes optimize project and building performances (Jia *et al.*, 2019). The digitalization, automation, and integration of the construction value chain also play a critical role in facilitating the digital transformation of the construction industry, enabling the construction industry to stay relevant in the era of the Fourth Industrial Revolution.

While the digital transformation of the construction industry driven by collaborative project delivery approaches can improve the performance of the industry, the nature of the construction industry gives rise to a general resistance toward adopting new technologies, resulting in laggardness in technology adoption (Hwang *et al.*, 2020; Sepasgozar *et al.*, 2016). However, there are only a limited amount of studies on the obstacles faced by practitioners when implementing collaborative project delivery approaches and the solutions that can promote the adoption of collaborative project delivery approaches in the context of the construction industry. Hence, this chapter answers the following questions:

(i) What are the obstacles faced by practitioners in adopting IDD?
(ii) How can the obstacles faced be overcome to increase the adoption of IDD?

To answer the research questions, this chapter has the following research objectives:

(i) To introduce the concept and technologies of IDD
(ii) To investigate the common obstacles faced by practitioners in adopting IDD
(iii) To propose feasible solutions to reap the benefits by increasing the implementation rate of IDD

While the chapter is focused on the context of the Singapore construction industry, the findings from this chapter can be extended to the global construction industry as the focus is placed on the underlying obstacles to implementing collaborative and integrated construction approaches and subsequently the solutions to overcome the obstacles. At the same time, rapid advancements in technologies need to be accompanied by collaborative and integrated construction approaches in order to overcome the challenges of low productivity, poor industry image, and fragmentation, which are common challenges across the global construction industry (World Economic Forum, 2018). Hence, the findings arising from this chapter are timely and essential to further drive the adoption of collaborative and integrated construction approaches and facilitate the digital transformation of the industry globally to stay competitive and relevant in the era of Fourth Industrial Revolution.

## Background

### *Concept of integrated digital delivery*

IDD was launched by Singapore's Building and Construction Authority to drive the digital transformation of the built environment value chain and improve the sustainability performance of the industry (Building and Construction Authority, 2018). IDD is supported by BIM, which refers to the digital representation of a building that can be used as a reliable shared information platform for decision-making throughout the building lifecycle (Azhar, 2011; Borrmann *et al.*, 2018). Beyond the digital representation of building models, IDD further encompasses other digitalized processes such as automated digital manufacturing and real-time monitoring processes. IDD has four phases that encompass the entire building lifecycle. In particular, the digital design phase occurs in tandem with the project initiation and design stages; the digital manufacturing and fabrication phase and the digital construction phase take place during the project construction stage; and the digital asset delivery and management phase occurs during the project handover and building operations and maintenance phases (Building and Construction Authority, 2018).

### *IDD across the building lifecycle*

#### *Digital design phase*

The digital design phase involves collaboration among various stakeholders to attain an optimized and coordinated design that meets the client,

**Table 1:** Digital design phase and its key technologies.

| Project phase | IDD phase | Key technologies |
|---|---|---|
| Initiation and design | Digital design | BIM<br>*Digital representation of a building with geometric and semantic information of the building components* |
| | | BIM-based sustainability analysis<br>*Sustainability performance analysis based on building design in BIM* |
| | | Constructability review and analysis<br>*Constructability analysis based on building design in BIM* |
| | | Augmented reality<br>*Technology to superimpose virtual information onto the user's physical environment* |
| | | Virtual reality<br>*Technology to create an interactive simulated experience based on the digital model* |
| | | Big Data analytics<br>*Technologies to analyze the huge volume, variety, and velocity of data to optimize decisions* |

*Source*: Adapted from Hwang *et al.* (2020).

regulatory, and downstream requirements (Building and Construction Authority, 2018; Hwang *et al.*, 2020). This phase will occur during the initiation and design phase of a standard project. In this phase, digital design and visualization tools are used to develop information-rich building models for easy visualization, while past project data are used to optimize building designs. The technologies also help to improve collaboration among stakeholders. Some of the key digital design and visualization tools include BIM and the various features for sustainability and constructability review and analysis, augmented and virtual reality, and Big Data analytics, as shown in Table 1 (Hwang *et al.*, 2020).

### *Digital manufacturing and fabrication phase*

The digital manufacturing and fabrication phase involves the establishment of a "Smart Factory" construction environment, so that digital designs can be translated into standardized components to be fabricated (Building and Construction Authority and Infocomm Media Development Authority, 2018; Hwang *et al.*, 2020). This phase will typically occur during the construction phase and helps to improve the quality of building components, reduces waste generation, and increases on-site productivity. As shown in Table 2, the key technologies in this phase include prefabrication, digital modeling and fabrication, robotics, digital ordering, central control, Radio Frequency Identification (RFID) material tracking, smart sensor network, and just-in-time delivery (Hwang *et al.*, 2020).

**Table 2:** Digital fabrication and manufacturing phase and its key technologies.

| Project phase | IDD phase | Key technologies |
|---|---|---|
| Construction | Digital manufacturing and fabrication | Prefabrication<br>*Production of components in a factory and transported on site for assembly* |
| | | RFID project material tracking<br>*Use of RFID tags to track project material in real time* |
| | | Digital ordering<br>*Procurement of individual components digitally* |
| | | Digital modeling and fabrication<br>*Modeling of individual components for fabrication digitally* |
| | | Central control<br>*Centralized monitoring and control of project activities* |
| | | Robotics<br>*Automated execution of pre-determined tasks to achieve set goals* |
| | | Smart sensor network<br>*Smart sensor technologies that track and monitor real-time operations* |
| | | Just-in-time delivery<br>*Delivery of components in the right amount to the right place at the right time* |

*Source*: Adapted from Hwang *et al.* (2020).

## *Digital construction phase*

The digital construction phase utilizes technologies that can conduct real-time monitoring and control to improve the operational efficiency of the project site and site safety, maximize productivity, and minimize reworks (Hwang *et al.*, 2020). Please refer to Table 3 for the key technologies in this phase, which include RFID tagged precast elements, virtual

**Table 3:** Digital construction phase and its key technologies.

| Project phase | IDD phase | Key technologies |
|---|---|---|
| Construction | Digital construction | 4D progress monitoring<br>*3D BIM model with a schedule dimension* |
| | | RFID tagged precast elements<br>*Use of RFID to track the production, delivery, and assembly of precast elements* |
| | | Virtual construction scheduling<br>*Simulation to identify out-of-sequence work processes and scheduling conflicts* |
| | | Digital QA/QC inspection<br>*Digital document management platform with checklists for inspection purposes* |
| | | Laser scanning<br>*Capturing of 3D geometry of surfaces to generate point clouds to develop a 3D model* |
| | | Virtual construction and collaboration<br>*Simulation of works required in a project to improve collaboration* |
| | | Co-located coordination<br>*Spatial intervention for stakeholders to collaborate and coordinate works* |
| | | Unmanned aerial vehicles<br>*Drones to reach locations inaccessible or dangerous for workers* |
| | | Photogrammetry<br>*Process images to reconstruct 3D objects* |
| | | BIM-to-field mobile applications<br>*Access to BIM on site using mobile devices* |

**Table 3:** (*Continued*)

| Project phase | IDD phase | Key technologies |
|---|---|---|
| | | Biometric controls and database<br>*Biometric scanners to record workers on site* |
| | | Real-time location services for monitoring<br>*Technologies to locate an asset or resource in real time for monitoring* |
| | | Smart crane navigation system<br>*Cranes that can automatically detect and position to place the precast element in its coordinates* |

*Source*: Adapted from Hwang *et al.* (2020).

construction scheduling and sequencing, virtual construction and collaboration, co-located coordination, digital quality assurance and quality control (QA/QC) inspection, 4D progress monitoring, laser scanning, photogrammetry, unmanned aerial vehicles, real-time location services for safety monitoring, BIM-to-field mobile applications, biometric controls and database, and smart crane navigation system (Hwang *et al.*, 2020).

### *Digital asset delivery and management phase*

Upon completion of the project construction phase, the information-rich building model is handed over to the building operations and maintenance team. The digital asset delivery and management phase occurs during the operations and maintenance phase of buildings and involves the use of technologies for real-time monitoring of the facility to enhance asset values (Hwang *et al.*, 2020). As shown in Table 4, the key technologies involved in this phase include BIM to manage defects and assets, digital progress reporting and claim, building management system, and smart facilities management technology (Hwang *et al.*, 2020).

## *Improvements in project performance*

The adoption of IDD is expected to improve project performance through the use of the key technologies identified. This is in alignment with

**Table 4:** Digital asset and delivery management phase and its key technologies.

| Project phase | IDD phase | Key technologies |
|---|---|---|
| Handover, operations, and maintenance | Digital asset delivery and management | BIM-based defect management system<br>*Identification and analysis of defects using BIM* |
| | | BIM for asset management<br>*Digital building assets data handed over to building owners through BIM* |
| | | Digital progress reporting and claim<br>*Tracking and monitoring of resources against planned progress that is linked to progress claim using a cloud platform* |
| | | Building management system<br>*Computer-based control system to control and monitor the building's equipment operations and performance* |
| | | Smart facilities management technology<br>*Use of sensors to monitor real-time operating data from different equipment within the building* |

*Source*: Adapted from Hwang *et al.* (2020).

previous studies where new technologies that are well integrated into an organization's work processes are presumed to improve performance (Loosemore, 2014; Sepasgozar *et al.*, 2016). Implementing IDD in projects has been perceived to improve overall project performance by an average of 5.15% (Hwang *et al.*, 2020). Specifically, project cost, quality, and schedule performances have been perceived to be improved by an average of 3.76%, 4.48%, and 4.44%, respectively (Hwang *et al.*, 2020).

The perceived improvements in project performance may be attributed to the mitigation of the poor information transfer environment of the construction site which enables accurate decision-making throughout the building lifecycle with the adoption of IDD (Hwang *et al.*, 2020). Improved information transfer among the stakeholders also minimizes unnecessary delays, on-site safety hazards and risks, reworks, and duplication of works, arising from miscommunications and poor scheduling, leading to increased productivity and improved project performance (Hwang *et al.*, 2020; Jia *et al.*, 2019).

The improvements in project performance were perceived to be greater by those with experience in IDD technologies, which is consistent with previous studies where experience in the technologies was found to influence the perceived usefulness and perceived ease of use of the technologies (Hwang *et al.*, 2020; Liu *et al.*, 2018; Venkatesh and Bala, 2008). This also reflects greater awareness of the benefits brought about by the IDD technologies when one has gained knowledge or experience in the technologies, which may further drive the adoption of IDD (Hwang *et al.*, 2020).

While IDD can improve project performance, the implementation of IDD in the construction industry remains in a nascent stage although a majority of the industry practitioners have experience with at least one of the IDD technologies (Hwang *et al.*, 2020). It is also important to note that the potential of the IDD technologies can only be maximized when complemented with the IDD project delivery approach, creating a synergy among the IDD technologies. Hence, it is necessary to understand the obstacles faced by practitioners in adopting IDD and the solutions to increase the adoption of IDD.

## Method and data presentation

the research process consists of four steps, utilizing a mixed research method, in particular, through a questionnaire survey and interviews. A literature review was first conducted to understand the concept and the technologies of IDD, obstacles faced by practitioners in adopting IDD, and the solutions to increase adoption of IDD. The identified obstacles and solutions and the corresponding categorization were validated in pilot interviews with industry experts. A questionnaire survey was then developed and distributed to target respondents to solicit the perceptions of industry practitioners on the significance of the obstacles and effectiveness of the solutions in adopting IDD. The data collected from the questionnaire survey were analyzed using nonparametric statistical tests. Following the data analysis, the findings were then validated through post-survey interviews. The post-survey interviews also enrich the findings of the study, providing a better understanding of the obstacles faced by practitioners and solutions that are perceived to be effective in promoting the adoption of IDD. Readers may refer to Hwang *et al.* (2020) for more details on the research methods and profiles of the respondents.

## Obstacles faced by practitioners in adopting IDD

The obstacles faced by practitioners in adopting IDD were investigated. 25 obstacles to IDD implementation have been identified and categorized into financial-related, government-related, management-related, project-related, knowledge-related, and client-related obstacles. Table 5 presents the mean, rank, and one-sample Wilcoxon-signed rank test results for the obstacles faced by practitioners in adopting IDD based on the survey responses. According to Table 5, all identified obstacles were found to be significant based on the mean rank analysis.

**Table 5:** Summary of results for the obstacles faced by practitioners in adopting IDD.

| Category | Code | Obstacle | Mean | Rank | One-sample Wilcoxon-signed rank test |
|---|---|---|---|---|---|
| Financial | O1 | High investment costs | 4.000 | 3 | 0.000 |
| | O2 | Slow recovery of investment | 3.790 | 8 | 0.000 |
| | O3 | Low profit margin | 3.613 | 16 | 0.000 |
| | O4 | Not economical as project scale is too small | 3.025 | 25 | 0.027 |
| | O5 | Budget overruns | 3.565 | 19 | 0.000 |
| | O6 | Uncertainty in cost benefits | 3.855 | 5 | 0.000 |
| Government | O7 | Lack of government support | 3.274 | 23 | 0.031 |
| | O8 | Lack of building standards and regulations for IDD | 3.581 | 18 | 0.000 |

**Table 5:** (*Continued*)

| **Category** | **Code** | **Obstacle** | **Mean** | **Rank** | **One-sample Wilcoxon-signed rank test** |
|---|---|---|---|---|---|
| Management | O9 | Lack of strategy to promote IDD | 3.484 | 21 | 0.001 |
| | O10 | Poor coordination between stakeholders | 3.774 | 10 | 0.000 |
| | O11 | Lack of measurement tools | 3.726 | 12 | 0.000 |
| | O12 | Lack of interest in IDD | 3.452 | 22 | 0.002 |
| | O13 | Costs of implementation outweigh benefits | 3.742 | 11 | 0.000 |
| | O14 | Inclination to stick to tried and tested methods | 4.065 | 1 | 0.000 |
| Project | O15 | Complexities of implementing IDD in projects | 4.016 | 2 | 0.000 |
| | O16 | More time and workload in pre-construction stage | 3.790 | 9 | 0.000 |
| | O17 | Unwillingness to implement IDD technologies not tried out by others in the industry | 3.719 | 13 | 0.000 |
| | O18 | Concerns with data security and protection | 3.177 | 24 | 0.164* |
| | O19 | Interoperability issues | 3.823 | 6 | 0.000 |
| | O20 | Prolonged approval procedures | 3.677 | 15 | 0.000 |
| Knowledge | O21 | Lack of knowledge in IDD standards and information on cloud and digital technologies | 3.823 | 7 | 0.000 |
| | O22 | Lack of relevant IDD expertise and experience | 3.871 | 4 | 0.000 |

(*Continued*)

**Table 5:** (*Continued*)

| Category | Code | Obstacle | Mean | Rank | One-sample Wilcoxon-signed rank test |
|---|---|---|---|---|---|
| Client | 023 | Lack of demand and interest in IDD from client | 3.694 | 14 | 0.000 |
| | 024 | Difficulty in performing the IDD requirements of the client | 3.597 | 17 | 0.000 |
| | 025 | Lack of standardized guidelines by the client | 3.565 | 20 | 0.000 |

## *Financial-related obstacles*

High investment costs were found to be one of the top critical obstacles faced by practitioners in adopting IDD. Costs of new technologies include hardware and software costs and the costs associated with training of existing staff and hiring of skilled personnel (Jiang *et al.*, 2018). Due to the transient nature of construction projects, the construction industry operates on a cost-driven structure. This has a strong influence on technological investments as construction organizations may perceive the huge upfront investment in technologies in the preliminary stages of adopting IDD as a huge deterrence (Enshassi *et al.*, 2016; Nagy *et al.*, 2018; Oesterreich and Teuteberg, 2016; Stewart *et al.*, 2004). In addition, small and medium enterprises (SMEs) make up a majority of the construction industry (Ngo *et al.*, 2020). SMEs are more vulnerable to variations in profit margins and have limited financial capabilities to invest in new technologies (Belle, 2017; Stewart *et al.*, 2004; Wong and Ng, 2017).

The implementation of new technologies entails the use of different work processes, which increases the risk of unanticipated costs, leading to a higher risk of budget overruns (Djokoto *et al.*, 2014; Harding, 2012). The uncertainty on the return on investment on IDD technologies further increases the perceived investment costs of new technologies (Oesterreich and Teuteberg, 2016). The same sentiment was also expressed by two post-survey interviewees, who stated that IDD implementation is perceived to require high investment costs due to the need to innovate processes to incorporate the new IDD technologies, re-train existing staff, and hire new talents competent in IDD. This finding is also in alignment with previous studies where high investment costs have been found to be one of the most significant obstacles faced by practitioners when adopting new technologies such as green building technologies (Darko and Chan, 2017; Hwang and Tan, 2012).

## *Government-related obstacles*

While digital technologies are readily available in the market, there are limited standards and guidelines for the implementation and management of the technologies in the construction industry. This can result in mishandling and underutilization of the technologies (Nagy *et al.*, 2018; Ngo *et al.*, 2020). Relevant standards and guidelines for the implementation and management of the technologies provide construction organizations with a systematic approach to implementing IDD technologies while ensuring a minimum standard of implementation to maximize the potential of the technologies. Furthermore, without government support in funding the implementation of new technologies, construction organizations, especially SMEs, may be reluctant to adopt new technologies (Enshassi *et al.*, 2016; Oesterreich and Teuteberg, 2016).

## *Management-related obstacles*

A critical obstacle to digital transformation is the intrinsic social resistance to embrace new technologies, especially in the construction industry, where people are inclined to stick to conventional methods (Enshassi *et al.*, 2016; Hewage *et al.*, 2008; Stewart *et al.*, 2004). Accordingly, the inclination to stick with tried and tested methods was found to be the key obstacle to IDD implementation as organizations may not be willing to take the risks of adopting new technologies or are not convinced with the benefits of using digital technologies (Rose and Manley, 2012). Furthermore, the construction industry is known to be a laggard in technology adoption and has a strong general resistance toward the adoption of new technologies (Oesterreich and Teuteberg, 2016; Sepasgozar *et al.*, 2016). One of the main reasons practitioners use proven technologies may be due to the potential liability that may arise from adopting new technologies and construction methods (Tatum, 1987). The inclination to stick to tried and tested methods may also be contributed by the schedule, costs, and quality pressures in any project, where there is little or no margin for experimentation (Liu *et al.*, 2018; Oesterreich and Teuteberg, 2019). The two post-survey interviewees agreed that one of the most critical obstacles faced when implementing new technologies was the general resistance toward change within the industry. In particular, it was highlighted that the demographics of the staff in the industry, where many of the staff are mature staff who are less attuned to the use of digital technologies in projects, may also have contributed to the inclination to stick with tried and tested methods in projects.

Furthermore, the top management may lack awareness of the strategic advantages of IDD and find that the costs of IDD implementation outweigh the anticipated benefits of IDD, resulting in a lack of interest in implementing IDD in projects (Belle, 2017; Enshassi *et al.*, 2016). In addition, the top management may not be able to assess the returns on investment in IDD technologies due to the lack of measurement tools (Oesterreich and Teuteberg, 2016). With the lack of interest in IDD, there is a lack of strategy to promote IDD within the organization, resulting in poor coordination between stakeholders (Hewage *et al.*, 2008; Nagy *et al.*, 2018; Oesterreich and Teuteberg, 2016; Stewart *et al.*, 2004).

## *Project-related obstacles*

The transient nature of construction projects also leads to unwillingness of the project teams to implement new technologies in projects as time and effort are required for the workers to understand how to operate the technologies (Enshassi *et al.*, 2016; Ngo *et al.*, 2020; Oesterreich and Teuteberg, 2016). Furthermore, the complexities of implementing IDD technologies in projects further increase the time and effort needed to understand and use the technologies (Mitropoulos and Tatum, 2002). When technologies are not used correctly, potential delays or reworks may occur (Stewart *et al.*, 2004). As such, project teams may not be willing to implement newly developed technologies not tried out by others in the industry (Enshassi *et al.*, 2016; Hewage *et al.*, 2008). The integration of new technologies into projects and buildings also transforms the way projects and buildings are designed, built, and managed, which increases the complexities of implementing IDD technologies into construction projects (Costa and Grilo, 2015). This requires changes to the skillsets required and the interactions within the interconnected value chain (Afuah and Bahram, 1995). Hence, the complexities of implementing IDD technologies in projects have been found to be one of the critical obstacles faced by practitioners in adopting IDD, as supported by the interviewees. The two interviewees also stated that IDD impacts multiple phases and requires extensive coordination, good planning, and execution with the multiple stakeholders involved, which further increase the complexities faced when integrating new technologies into projects.

In addition, when implementing new construction technologies, longer approval procedures may be required, further deterring the project team from implementing new technologies due to the time constraints of a project (Eggers *et al.*, 2018; Ngo *et al.*, 2020). Apart from that, the project team is made up of multiple stakeholders that operate on various software, which can lead to interoperability issues when using a new technology (Enshassi *et al.*, 2016; Stewart *et al.*, 2004). Interoperability issues and varying levels of competencies in managing the IDD technologies across the construction value chain also increase the complexities of IDD implementation in projects (Oesterreich and Teuteberg, 2016). The unfamiliarity and low level of knowledge of IDD and its technologies also increase the perceived complexities of IDD implementation in projects as users are uncertain of the problems that may be faced during the implementation process and how to overcome them (Adriaanse *et al.*, 2010). With the increasing data volume and demand for information sharing through cloud technologies, there are concerns over data security due to the presence of sensitive information within the organizations (Nagy *et al.*, 2018; Oesterreich and Teuteberg, 2016; Stewart *et al.*, 2004). However, concerns with data security and protection were found to be statistically insignificant based on the one-sample Wilcoxon signed-rank test. This may suggest a poor understanding of the significance of data security in the construction industry (Gu and London, 2010). This could reflect the limited level of awareness, knowledge, and experience in IDD and the technologies that can be deployed in projects possessed by the practitioners (Acquisti *et al.*, 2017). Furthermore, the existing collaboration and technology practices do not support the adoption of collaborative work practices, which may also contribute to the lack of awareness in data security requirements (Alreshidi *et al.*, 2018).

## *Knowledge-related obstacles*

To adopt IDD in construction projects, employees must have some level of knowledge about the IDD standards and technologies (Nagy *et al.*, 2018; Oesterreich and Teuteberg, 2016). However, the existing level of IDD competency and experience of workers in implementing IDD technologies in projects is relatively low, which may hinder construction organizations from implementing IDD due to the need to send the workers for training (Enshassi *et al.*, 2016; Hewage *et al.*, 2008). Furthermore, the level of competency of workers in the technologies across the value chain varies greatly (Oesterreich and Teuteberg, 2016). The transient nature of the projects further limits the experimentation of new technologies and may deter organizations from sending their employees for training.

## *Client-related obstacles*

As clients typically determine the project requirements, the lack of interest in IDD from the client has a significant impact on the implementation of IDD in projects (Belle, 2017; Enshassi *et al.*, 2016; Hewage *et al.*, 2008). Furthermore, clients have several priorities to meet and usually favor the lowest price bidders (Ngo *et al.*, 2020). Hence, to keep the tender prices competitive, contractors without the technological expertise tend to not implement new technologies (Ngo *et al.*, 2020). As the technological approach to construction is relatively new to the construction industry, there may also be lapses in the guidelines set by the clients during the tendering stage, resulting in difficulties for contractors to fulfill the IDD requirements of the client (Hewage *et al.*, 2008; Nagy *et al.*, 2018; Oesterreich and Teuteberg, 2016; Stewart *et al.*, 2004). This further deters organizations from adopting IDD, as high uncertainties may lead to higher project risks and costs.

## Solutions to increase adoption of integrated digital delivery

To overcome the obstacles faced by practitioners to increase the adoption of IDD, 18 solutions from across the government, organization, and industry approaches have been identified. Table 6 shows the mean, rank and one-sample Wilcoxon-signed rank test results for the solutions perceived to be effective by practitioners to overcome the obstacles faced in adopting IDD based on the survey responses. According to Table 6, all the identified solutions were perceived to be effective in promoting the adoption of IDD based on the mean, rank, and one-sample Wilcoxon-signed rank test results.

**Table 6:** Summary of results of solutions to overcome the obstacles faced.

| Code | Solutions | Target obstacles | Mean | Rank | One-sample Wilcoxon-signed rank test |
|---|---|---|---|---|---|
| S1 | Provision of financial incentives | O1–O7 | 4.161 | 2 | 0.000 |
| S2 | Development of building solutions and standards for IDD implementation | O8 | 3.903 | 12 | 0.000 |
| S3 | Adoption of IDD in public projects | O14, O17 | 4.194 | 1 | 0.000 |
| S4 | Monetary subsidies | O21, O22 | 4.065 | 8 | 0.000 |
| S5 | Promote research and development in IDD technologies | O21–O23 | 4.000 | 9 | 0.000 |
| S6 | Enforcement | O12 | 3.952 | 11 | 0.000 |
| S7 | Engage IDD skilled professionals | O10 | 4.000 | 9 | 0.000 |

(*Continued*)

**Table 6:** (*Continued*)

| Code | Solutions | Target obstacles | Mean | Rank | One-sample Wilcoxon-signed rank test |
|---|---|---|---|---|---|
| S8 | Development of IDD framework | O9, O11 | 3.903 | 12 | 0.000 |
| S9 | Front-end planning | O2, O5, O6 | 4.097 | 5 | 0.000 |
| S10 | Effective communication between key stakeholders | O10 | 3.903 | 12 | 0.000 |
| S11 | Perform studies and analyses on project requirements and complexities in advance | O13, O16, O18, O19 | 3.839 | 15 | 0.000 |
| S12 | Training | O15–O22 | 4.129 | 4 | 0.000 |
| S13 | Introduction of IDD knowledge in curriculum | O21 | 4.081 | 7 | 0.000 |
| S14 | Lower interest rates for IDD projects | O1–O6 | 3.484 | 18 | 0.000 |
| S15 | Competitive advantage given to organizations implementing IDD in tenders | O12, O14, O17 | 3.677 | 16 | 0.000 |
| S16 | Pilot program and sharing of successful projects | O15, O22, O23 | 3.677 | 16 | 0.000 |
| S17 | Client to advocate for IDD | O23 | 4.097 | 5 | 0.000 |
| S18 | Client to provide clear procedures to fulfill IDD requirements during the tendering stage | O15, O24, O25 | 4.145 | 3 | 0.000 |

### Solutions to overcome finance-related obstacles

Provision of financial incentives and monetary subsidies is the most straightforward solution to promoting IDD implementation by assisting the organizations in overcoming their financial obstacles (Chan *et al.*, 2017; Stewart *et al.*, 2004; Suprun and Stewart, 2015). Financial obstacles are one of the most critical obstacles to technology implementation, especially for SMEs (Wong and Ng, 2017). As such, the provision of financial incentives to offset the upfront investment costs and monetary subsidies for training costs will help to overcome the financial obstacles and promote IDD implementation. Apart from the financial incentives provided by the government, industry support is also required to promote IDD implementation. For instance, lower interest rates can be offered by banks for projects implementing IDD to increase the price competitiveness of these organizations (Abdul-Aziz, 1994; Yik *et al.*, 2006). This can help to partially offset some of the costs incurred by organizations when adopting IDD. Organizations should also perform sufficient studies and analyses on project requirements and complexities in advance so that front-end planning can be conducted to prevent budget overruns and prepare for uncertainties arising from the adoption of new processes (Lunn and Stephenson, 2000; Smith, 2014).

### Solutions to overcome government-related obstacles

National leadership is a critical factor in driving technological implementation in the construction industry (Ngo *et al.*, 2020). In particular, the government should set standards for the industry to ensure that a minimum standard of implementation can be met and to provide guidelines as a reference for the industry to adopt best practices and maximize the potential of the IDD technologies (Blayse and Manley, 2004; Smith, 2014; Stewart *et al.*, 2004). This helps to overcome the obstacles related to the lack of standards and guidelines for the implementation and management of the technologies in the construction industry.

### Solutions to overcome management-related obstacles

Management-related obstacles may be overcome through the adoption of IDD in public projects to spur demand in IDD and this can demonstrate the potential of IDD in improving the performances of the projects the organizations are involved in (Smith, 2014). These projects can also serve as references for others in the industry to implement IDD. Demand in IDD may also be spurred when the government mandates and enforces its use in projects (Blayse and Manley, 2004; Chan *et al.*, 2017). Alternatively, the comparative advantage given to organizations that propose to implement IDD in projects during the tendering processes may also spur IDD demand (Abdul-Aziz, 1994; Yik *et al.*, 2006). To adopt IDD in projects, construction organizations need to provide training for staff and workers to be knowledgeable and skilled in IDD (Oesterreich and Teuteberg, 2016; Suprun and Stewart, 2015). Organizations can also hire or engage IDD professionals to guide the IDD implementation and overcome the potential issues with coordination between stakeholders (Blayse and Manley, 2004). Effective communication is essential between the key players in the IDD implementation process so that the entire team can progress in the right direction and embrace the use of technologies in the projects (Smith, 2014). At the same time, organizations must develop an IDD framework to ensure that the organization processes are suitable for IDD implementation and to guide the organization in IDD implementation (Lunn and Stephenson, 2000; Smith, 2014; Stewart *et al.*, 2004). The IDD framework can serve as a measurement tool and provide a strategy to promote IDD adoption for the organization.

## *Solutions to overcome project-related obstacles*

To overcome project-related obstacles, several solutions across government, industry, and organization approaches may be utilized. For instance, the adoption of IDD in public projects allows stakeholders to familiarize themselves with the IDD technologies and delivery approach (Smith, 2014). Apart from that, comparative advantages given to organizations that propose to implement IDD in projects during the tendering processes may also encourage adoption of IDD to improve the competitiveness of the organizations, reducing the unwillingness to implement IDD. However, front-end planning and sufficient studies and analyses need to be conducted in advance to prepare for the changes in work processes and complexities that may arise from IDD implementation (Lunn and Stephenson, 2000; Smith, 2014). To help organizations conduct front-end planning, pilot programs and the promotion of past successful IDD projects can be shared with existing practitioners to showcase the benefits of IDD and provide a reference for others looking to implement IDD into projects (Blayse and Manley, 2004; Smith, 2014; Stewart *et al.*, 2004; Suprun and Stewart, 2015). At the same time, clients should provide clear procedures to fulfill IDD requirements to contractors during the tendering process so that the contractors are aware of the expectations and can plan for the use of the technologies (Blayse and Manley, 2004; Lunn and Stephenson, 2000; Mitropoulos and Tatum, 2002). IDD implementation also requires skilled professionals to manage the use of IDD in projects. Hence, it is also essential for construction organizations to provide training for staff and workers to be knowledgeable and skilled in IDD (Oesterreich and Teuteberg, 2016; Suprun and Stewart, 2015).

## *Solutions to overcome knowledge-related obstacles*

Knowledge-related obstacles may be overcome with the government providing monetary subsidies for training courses. In addition, integrating knowledge on IDD technologies and processes into the curriculum provided by educational institutions can play a key role in preparing the industry for IDD implementation. This equips the future workforce with proficiency in handling and managing IDD in projects (Smith, 2014; Suprun and Stewart, 2015). Knowledge of IDD technologies and management of the technologies can also be shared with existing practitioners through pilot programs and the promotion of past successful IDD projects (Ngo *et al.*, 2020). Promotion of research and development in IDD technologies can also help to increase the IDD knowledge of the entire value chain.

## *Solutions to overcome client-related obstacles*

As clients play a key role in driving demand of new technologies in projects, encouraging clients to advocate for IDD can drive the adoption of IDD. This may be done through the promotion of research and development in IDD technologies, where the advancement of such technologies provides huge potential to improve the performance of the projects. In addition, sharing of successful case studies to the clients can demonstrate the potential benefits that can be derived from IDD adoption, further spurring the demand of IDD by clients (Blayse and Manley, 2004; Smith, 2014; Stewart *et al.*, 2004; Suprun and Stewart, 2015). Besides encouraging clients to advocate for IDD, it is also important for clients to provide clear procedures to fulfill IDD requirements to contractors during the tendering process so that the difficulty in performing the IDD requirements of client and lack of standardized guidelines by the client can be mitigated, further encouraging IDD adoption (Blayse and Manley, 2004; Lunn and Stephenson, 2000; Mitropoulos and Tatum, 2002).

## *Key solutions to overcome obstacles*

The three most effective solutions were found to be adoption of IDD in public projects, provision of financial incentives, and client to provide clear procedures to fulfill IDD requirements during the tendering stage.

Adoption of IDD in public projects was found to be the most effective solution. The adoption of IDD in public projects spurs demand in IDD technologies and allows contractors to be familiarized with the use of IDD in projects. This helps to increase the IDD proficiency of contractors and can encourage implementation of IDD in future projects (Liu *et al.*, 2018). At the same time, the increased demand of IDD technologies may further reduce the costs of IDD technologies due to economies of scale, mitigating the obstacle of high investment costs. Furthermore, these projects can serve as guidelines for construction organizations and showcase the benefits of implementing IDD, which can mitigate the key obstacle of the inclination to stick to tried and tested methods and uncertainty in cost benefits (Smith, 2014; Stewart *et al.*, 2004). With the adoption of IDD in public projects, these public projects demonstrate how owners, developers, consultants, and contractors can set specific goals in IDD projects and how to utilize IDD to achieve the set goals, reinforcing the introduction of IDD into the construction ecosystem (Smith, 2014). Both interviewees also agreed that the adoption of IDD in public projects is effective in promoting the adoption of IDD. It was highlighted that while involvement in public IDD projects requires organizations to be more stringent when managing the construction program, the projects serve as platforms to train their employees in IDD and to develop best practices and knowledge of IDD. At the same time, consultants and contractors must work together to develop feasible solutions for integrating the IDD technologies into the projects, which can be carried forward to future projects.

Provision of financial incentives was found to be the second most effective solution to promoting the adoption of IDD. This is in alignment with previous studies by Chan *et al.* (2017), Stewart *et al.* (2004), and Suprun and Stewart (2015), who found that financial incentives are essential in the promotion of technological developments and uptake in the construction industry. Furthermore, high investment costs were one of the most significant obstacles to IDD implementation. Providing financial incentives can help organizations partially offset the costs associated with technology implementation, reducing the financial risks to be undertaken by organizations during the process of IDD implementation. This may be especially effective in Singapore's construction industry, where a majority of the projects are awarded to the lowest price bidder, although there is a shift toward the price quality evaluation method. While both interviewees agreed that financial incentives are important tools to encourage IDD implementation, post-survey interviewee A stated that organizations must not be dependent on financial incentives as it is not a long-term solution to develop continuous innovation in the organization. This requires a mindset change in the industry, which can be achieved through the adoption of IDD in public projects and pilot programs and sharing of successful projects to educate the wider industry, including the clients and end users, on the benefits of the IDD technologies and IDD approach to construction to increase the demand for IDD. This may be done through organized tours to facilities constructed using the IDD approach for building owners and potential end users (Hwang, 2018). Alternatively, the use of a virtual walkthrough in construction conventions and exhibitions can also allow the public to appreciate how IDD technologies and approaches can be utilized in buildings (Hwang, 2018).

The third most effective solution to promoting adoption of IDD was found to be the provision of clear procedures to fulfill IDD requirements during the tendering stage by the client. This can help to mitigate the complexities of implementing IDD technologies into the project, which was found to be the second key obstacle to IDD implementation — difficulty in performing IDD requirements of client and lack of standardized guidelines from the client. By providing clear procedures to fulfill IDD requirements during the tendering stage, contractors can make better informed decisions and planning to incorporate technologies into projects (Chan *et al.*, 2017). Informed planning can be critical in IDD projects as IDD requires collaboration and strategic alignment among all stakeholders along the value chain in order to be utilized effectively. The contractor may also need to engage with skilled IDD professionals to guide the team in fulfilling the IDD requirements, mitigating the lack of relevant IDD expertise and experience of the project team. Consequently, the risks that the contractor takes on in the project can be managed. This can also help to reduce the high investment costs in new technologies, in particular, the contingency provision for uncertainties arising from the use of new technologies in projects.

## Conclusion

Integrated and collaborative project delivery approaches are increasingly being introduced and emphasized to integrate the fragmented construction value chain and complement the use of rapidly advancing technologies. The integration of the value chain is expected to improve the performances of the building and construction industries. With a focus on IDD, the obstacles faced by practitioners in adopting IDD and solutions to increase IDD adoption were investigated. It was found that the top three obstacles to IDD implementation are inclination to stick with tried and tested methods, complexities of implementing IDD technologies into projects, and high investment costs. To overcome the obstacles and promote the adoption of IDD, the top three solutions were found to be adoption of IDD in public projects, provision of financial incentives, and clear information on procedures to fulfill IDD requirements during the tendering stage. By understanding the key obstacles and solutions to IDD implementation, governments and the global construction industry can better facilitate the digital transformation of the construction industry through the use of integrated and collaborative project delivery approaches, and in turn improve information transfer and sustainability performance of the construction industry. The findings can also be extended to the global construction industry as the underlying obstacles to implementing new technologies and collaborative project delivery approaches could be similar, albeit at varying degrees. These findings also contribute to practice, serving as a reference to governments when developing roadmaps to promote adoption of integrated and collaborative construction approaches. This can help drive the digitalization and automation of the construction value chain to enable smart and sustainable buildings throughout its lifecycle.

## Acknowledgments

This research is supported by the National University of Singapore, Office of the Deputy President (Research and Technology), HSS Seed Fund.

# About the Authors

**Jasmine Ngo** is a Ph.D. candidate in the Department of Building, National University of Singapore. Her research interests are in the areas of project and technology management and technological competency. She can be contacted at jasminengo@u.nus.edu.

**Bon-Gang Hwang**, Ph.D., is a professor in the Department of Building, National University of Singapore. He holds a PhD from the University of Texas, Austin. His research interests are in the areas of sustainable construction project management, project and technology management, performance analytics and improvement, and risk management. He can be contacted at bdghbg@nus.edu.sg.

**Wan Yi, Priscilla, Her** has graduated from the Department of Building, National University of Singapore. Her research interests are in the areas of sustainable construction project management and project and technology management. She can be contacted at priscillaher@u.nus.edu.

# Chapter 3

# The Relationship between Project Delivery Method and Project Work Environment and Collaboration

Edward Minchin

*University of Florida*

## Abstract

Almost anyone who has worked daily on a construction site or field office knows the feeling of arriving on site for another day of work, and seeing the vehicle of someone that works for one of the other parties of the construction contract, and having their stomach muscles tighten in anticipation of the confrontation that seems to happen every day. There may not be a particular issue that appears destined to result in this seemingly inevitable conflict, but the working environment on the project is so poor that any little disagreement or inconvenience will almost surely escalate into a major confrontation. There are many possible factors at work here: personalities, prejudices, third parties (such as utilities), constructability issues, etc. One of the most subtle, yet persistent factors, is the delivery system being employed on the project. Different delivery systems do create different work environments and levels of collaboration, but the single biggest contributor to stress in the construction workplace is whether the contractor feels that there is enough money in the contract to ensure that, with proper execution of the work and barring a major calamity, the company will emerge from the project with a profit — or at least without a loss.

## Background

The history of the building professions known today as "the contractor" and "the designer" extends back to the world's first civilizations. The "master builder" was a respected member of any community from the time of the ancient Mesopotamians up to the early 20th century. The earliest building codes discovered date back to approximately 1772 BCE, when the Babylonian emperor Hammurabi enacted his famous set of codes. Some of Hammurabi's codes — or laws — pertained to construction, including the following:

- 229 — If a builder builds a house for someone, and does not construct it properly, and the house which he built falls in and kills its owner, then that builder shall be put to death.
- 230 — If it kills the son of the owner, the son of that builder shall be put to death.
- 231 — If it kills a slave of the owner, then he shall pay, slave for slave, to the owner of the house (Hammurabi, 2020).

The severity of these consequences for poor construction must surely be an inverse indication of how seriously Babylonian builders took their work and how highly their successful work was regarded.

The Apostle Paul, writing in the Christian Bible, states that "According to the grace of God which is given unto me, as a wise masterbuilder, I have laid the foundation, and another buildeth thereon. But let every man take heed how he buildeth thereupon" (Bible, 1611). Here, as in every ancient attestation, the master builder's profession is described as one which entails both design and construction, along with wisdom and knowledge.

For millennia, the master builder held the preeminent position as the professional responsible for the built environment. He was a "one-stop shop" for getting something built. One simply had to tell him what was desired, whether it was a building, a road, a bridge, or a dam, and he would both design and build it. He was a true turnkey contractor. However, this status began to erode in the Renaissance period, when artists began taking on architectural commissions, despite having absolutely no understanding of the building arts or structural mechanics. This disruption of the profession set into motion a series of events that, 400 years later, culminated in the astonishing disappearance of the master builder in the 20th century.

The move away from the master builder tradition started in vertical, or building, construction, when the late nineteenth century saw the emergence of two new influential figures: the architect and the contractor. The increasing preeminence of the architect was largely attributable to Sir John Soane, the father of modern architecture who rose to fame and influence at the beginning of the nineteenth century. Although he himself was the son of a contractor (bricklayer), Soane arrogantly insisted that the architect should completely separate himself from the vulgar pursuit of building and from the non-professionals engaged in it (Bradbury, 2018). The general contractor, on the other hand, was someone without the desire or ability to design, but who was wholly engaged in the process of construction. Still, in spite of being displaced from vertical construction by the architect and the contractor, the master builder entered the twentieth century as the predominant delivery agent for highway, bridge, and road building. As such, the master builder was a generalist who might design a road, a bridge, or a dam and do his own soil analysis and design, hydrological analysis, hydraulics design, sewer design, structural analysis and design, and whatever else a project might require. Beginning in the 1930s, though, a movement toward specialization was stirring. The emerging specialists — structural engineers, geotechnical engineers, hydraulics engineers, drainage engineers, traffic engineers, and roadway design engineers — all brought increasing levels of expertise and sophistication to the many elements of roadway construction. By mid-century, with the legendary series of federally funded road tests conducted by the American Association of State Highway Officials (AASHO) — largely for the construction of what would become Interstate Highway 80 in Illinois — the Age of Specialization had fully dawned in the transportation construction industry (see Figure 1). The change had happened roughly two decades earlier in the vertical construction industry. The American Institute of Architects (AIA) established a Design–Bid–Build (DBB) Task Force in 1975 and first allowed their members to work in a Design–Build (DB) contract in 1978. They established their first DB contracts in 1985, and in

**Figure 1:** Historic marker on Interstate Highway 80 in Illinois identifies the site of some of the AASHO Road Tests.

1991, the AIA established a policy in favor of using DB in the public sector (AIA, 2008).

This new age brought about the widespread adoption of a new construction project delivery system — the DBB delivery system — along with the accompanying unit price contract and the competitive (low-bid) procurement system.

## Design–bid–build

US public transportation agencies had first used the DBB delivery system for highway construction in the 1930s, though it had been used sporadically in vertical construction in the preceding decades. Also known as the "linear" or "traditional" system, this delivery method is characterized by a series of "silos", where it is rare for the designer of the project to ever meet the constructor of that project. Design and contract documents are completed and stored until the job is "let". The time between the completion of design and the letting of a public project can be up to seven years and even longer for a private project. Figure 2 shows the timeline of a typical DBB project.

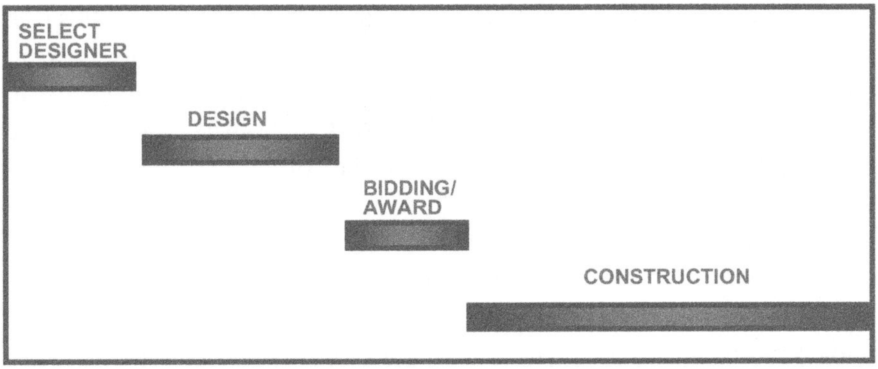


**Figure 2:** Relative Timeline for the Design-Bid-Build Process.

DBB is still the most popular delivery system in the US because it often results in the lowest price (not cost), is a relatively transparent system, is popular with contractors, is a familiar system with well-tested legal parameters, and is simple in execution. A small project lacking complexity is well suited for DBB.

Because the DBB delivery system methodically isolates each project activity from the others, it almost always requires the longest duration from conception to completion of any delivery system. The method began to lose favor as life sped up and motorists grew tired of contractors slowing them down on roadways that seemed interminably under construction. As a result, voters began demanding that new roads be built faster.

Another negative and expensive feature of the DBB delivery system is that it is arranged so as to perpetuate adversarial relationships between project parties before construction ever begins. Indeed, the very organization of a DBB project sets the parties against one another. It is therefore not surprising that the DBB system generates more litigation than any other construction delivery system does. Details of this aspect of DBB will be explored in the last section of this chapter.

## Design–build

Speed was the watchword in the 1980s, when project teams began to move toward fast-track, or integrated, construction project delivery. The first of the modern integrated construction project delivery methods to gain widespread popularity was DB, having become common by the 1990s (the author participated in a $50,000,000 Pilot Study of DB for highway work conducted by the Florida Department of Transportation that began in 1991). According to the Design-Build Institute of America, the name "Design–Build" was settled on in 1993 as the "official name for the integrated project delivery method their [member] companies had been using more frequently than the traditional low-bid approach" (DBIA, 2020).

Like any fast-track delivery system, DB offers the advantage of speed, but there are other advantages to using this system instead of DBB in numerous situations. The "integrated" part of integrated project delivery means that the constructor, i.e., the contractor, participates in the design of the project from the beginning. Generally, the owner provides a set of plans to potential design-builders, laying out the project details in the most basic terms. These plans are non-obligatory as long as no material changes are made. For example, if a set of conceptual plans calls for a four-story building with brick siding and a bathroom on each floor, it may be acceptable for the design-builder to develop a detailed design that calls for a four-story building with a glass-and-steel exterior and two bathrooms on the top floor — the four stories being expected, but the details non-obligatory. In addition to saving time and money, the presence of contractor representatives on the design team can also add quality, if they can help formulate design innovations that make the most of their particular workers, equipment, and expertise. This added value is not a given, however. While research conclusively shows that DB shortens project duration compared to DBB, the results are inconclusive as to whether DB increases quality or saves money (Ibbs *et al.*, 2003). Figure 3 shows the relative timeline for a DB project.

From the owner perspective, the "single point of contact" aspect of DB is very attractive. The convenience of only having to contact one entity to discuss project issues is obvious, but this convenience has its costs. The comparison Figure 4 presents of the organizational charts for DB and DBB shows that the price an owner pays for this single point of contact is that the design professional (DP) on a DB project, in effect, now works for the contractor in the majority of cases. While some DB contracts explicitly name the architect as the leader, many are silent on who leads. When this is the case, particularly on a large DB joint venture, the economic and financial strength of the contractor organization gives it a

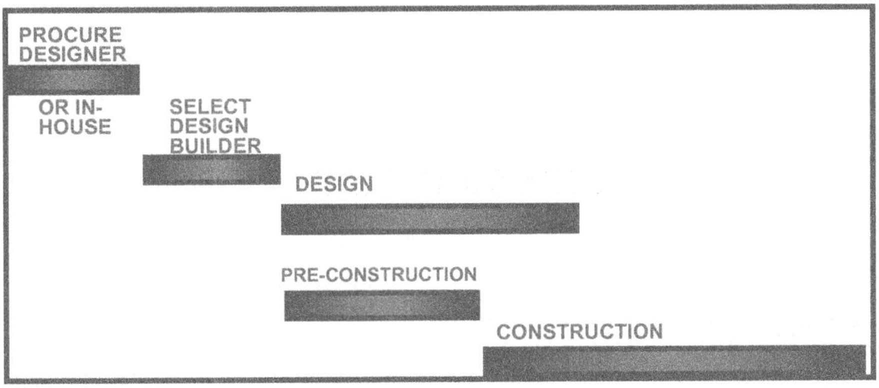


**Figure 3:** Relative Timeline for the Design-Build Process.

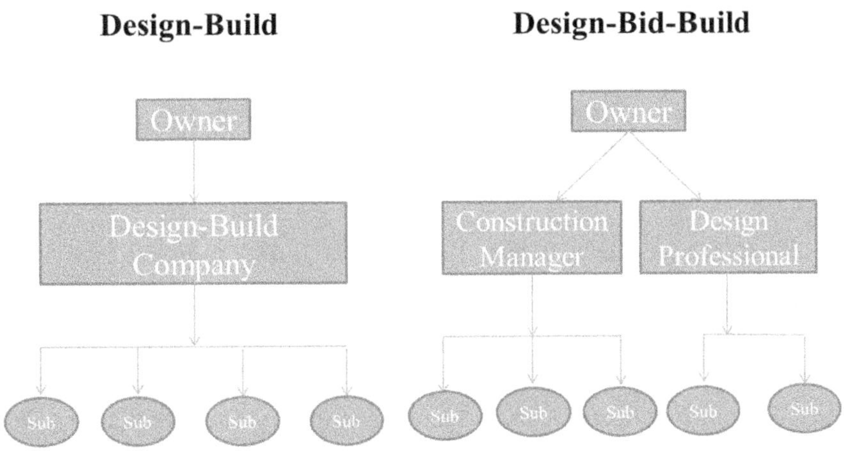


**Figure 4:** Organizational Charts for Design-Build, Left, and Design-Bid-Build, Right.

much greater capacity to procure surety bonds, which in turn propels it to the leadership position. So, on the largest vertical projects and on virtually all horizontal and industrial DB projects, the contractor leads the DP (whether an architect or an engineer). In these cases, the owner effectively loses control over the design process, and the DP has a vested interest in siding with the contractor in any dispute between the owner and the contractor. This built-in partiality is clearly a disadvantage to the owner. For many years, the Federal Highway Administration (FHWA) evinced its awareness of this drawback to the DB method — its fear of letting "the fox guard the henhouse" — in its prohibition of federal funding of DB projects. In this view, the entity that had always protected the owner's interests was now on the other side of the negotiating table. Over the years, well-established owners with reliably expansive project portfolios have been largely able to sidestep this problem, since the threat of removal from such an owner's approved vendors' list acts to deter any potentially fraudulent DP. However, for a small owner, the threat of the DP siding with a contractor to the owner's detriment remains real.

Innovation is another important advantage of DB. A good design-builder can take a set of conceptual plans and generate a design containing positive elements that the owner had not even thought of. One particular state department of transportation (DOT) has published several case studies of such successfully innovative redesigns. Based on this history of successful innovation, this DOT now lists it as one of its criterion for using DB on projects as "Opportunity for Innovation". In choosing the design-builders for its DB projects, it counts the technical quality parameter for a complex project as 25–30% of the total metric. For DB projects with little flexibility, it only counts this parameter as 5–15% of the total metric.

## Three case studies from the same state

One project entailed the widening of 6.2 miles of the interstate highway in an urban environment. Since the widening would all take place in the median, this team's innovation centered on a plan to incorporate two conveyor belt systems to transport stone and asphalt to the interstate median during construction, under traffic. The conveyor belt system began in the southbound right-of-way (ROW) and rose on a temporary structure that passed over the southbound lanes and into the median. This eliminated the need for large dump trucks to decelerate and execute a left turn into the median. Once executed, this design eliminated 237 lane closures and 11,800 delivery truck trips in and out of the median area.

The second project called for the widening of 6.8 miles of the rural interstate highway. Again, the widening would all take place in the median. The DB team awarded the project proposed an unheard-of temporary access bridge and ramp to minimize the use of the highway for materials hauling and worker travel. The results were similar to those of the first case study.

**Table 1:** Results of Innovations on Three D-B Projects in the Same State.

| Project Description | Innovation Description | Savings ($M) |
|---|---|---|
| 6.2-mi. Urban Interstate Highway Widening Project | Conveyor Belt system to get materials into the median from trucks in right ROW | 1.5 |
| 6.8-mi. Rural Interstate Highway Widening Project | Bridge to allow trucks to dump loads in median, then cross to right side for return | 5.0 |
| Interstate Interchange | Replaced four-level stacked interchange with a two-level turbine interchange | 30.0 |

The DB team awarded the third project won the bid partially because it proposed an innovative conversion of a four-level stacked interchange to a two-level turbine interchange. This single innovation eliminated approximately 2,000,000 cubic yards of borrow and obviated the need for a highway detour that would have required the setting of steel girders during the four-level construction. Because the turbine interchange lowered the roadway embankment height by 40 feet, it eliminated any lane closures that might have been needed during icy conditions on a four-level interchange. This innovation also lowered maintenance and future construction costs for the life cycle of the structure. Table 1 shows the savings from the innovations in each case.

Another way that DB is an improvement on DBB is that the system's arrangement naturally engenders less animosity and confrontation among the contract parties. While exceptions can certainly be cited, research and experience back this assertion. Details of this aspect of DB will be discussed further in the last section of this chapter.

## Public–private partnerships

While the term "public–private partnership" (P3) has been variously defined, this chapter will use the US Federal Highway Administration (FHWA) definition:

> Contractual arrangements between public and private sector entities pursuant to which the private sector is involved in multiple elements of

> public infrastructure projects. Unlike conventional methods of contracting for a project, in which discrete functions are divided and procured through separate solicitations, [P3s] contemplate a single private entity being responsible and financially liable for performing all or a significant number of functions in connection with a project. The "private partner" is typically a consortium of private companies with expertise in the different functions to be performed (design, construction, financing, operation and/or maintenance). In transferring responsibility and risk for multiple project elements to the private partner, the procuring agency shifts certain risks to the private partner and focuses on desired outcomes instead of detailed project specifications. The private partner receives the opportunity to earn a financial return commensurate with the risks it assumes. Structured in multiple forms, [P3s] vary generally according to the scope of responsibility and degree of risk assumed by the private partner with respect to the project. In each case the private partner assumes financial risk in some form — for example, through an equity investment, liability for indebtedness, a fixed priced contract, or a combination thereof — and risks related to project design, construction, operation and maintenance, as applicable (FHWA, 2016).

Figure 5 shows the organizational chart for a P3 project.

The crumbling of the infrastructure in the US and around the world is nearing a point of crisis, and public agencies do not have the resources to

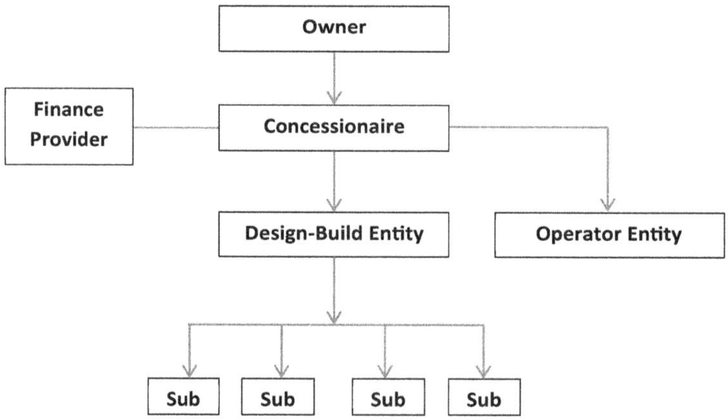


**Figure 5:** Organizational Chart for a Public-Private-Partnership Project.

keep up with the demand for new facilities, much less make up ground and catch up to the problem. Whether owner- or constructor-initiated, P3s offer cash-strapped agencies a way to get a large project, or project portfolio, built at almost no cost to the public. So far, most completed P3 projects have been very large "mega-projects".

At the core of almost every P3 project lies the DB delivery system. Because certain contracting schemes that have emerged from the DB system can be seen as specific forms of P3 contractual arrangements, questions often arise about how they each differ from a P3 contract. Examples of these kinds of contracts are BOT (build–operate–transfer), BOOT (build–own–operate–transfer), DBFO (design–build–finance–operate), and DBFOM (design–build–finance–operate–maintain). The confusion around these terms comes from their general lack of fixed definitions and the way their usage varies regionally. International comparisons have shown that "one man's BOOT is another's DBFO" (USDOT, 2009). Even as these acronyms attest to the integration of the P3 and DB schemes, it is clear that the many financial arrangements that define P3 are also harbingers of differences between the "straight DB" contract and P3.

Indeed, the presence of DB contracts at the core of P3 projects gives them many of the advantages inherent to DB. However, while the financial arrangements of P3 generally allow projects to be built in the first place, they vitiate and even eliminate some of these advantages. The concessionaire (sometimes called the developer or special-purpose vehicle) is the agent delivering the two primary functions that differentiate DB contracts from P3 contracts: arranging for project financing, typically through a combination of equity and debt; contracting with an operations/maintenance provider to preserve the facility's serviceability during the concession term. See Figure 5 for the position of the concessionaire in the P3 organizational chart.

In the United States, P3 contracting to date has been used almost entirely on heavy transportation facilities such as highways, bridges, and tunnels. Despite the extensive global use of this delivery approach for social infrastructure and non-highway transportation facilities, P3s in these sectors are still in their infancy in the US. Moreover, in spite of the use of P3 globally for many types of construction other than heavy transportation systems, it stands as the dominant project type for highways, bridges, and tunnel construction around the world (Mahmudova, 2013).

As discussed, innovation is one of the most important advantages offered by any integrated delivery system, and since almost all P3 projects are operated from a DB core, one would expect P3 to include innovation as one of its many advantages. Leiringer (2006) made a key research contribution in assessing the influence of P3s on innovation, focusing particularly on technological innovations in the design and construction phases of P3 projects. This assessment identified four innovation drivers often attributed to the P3 system by proponents: (1) increased collaboration; (2) greater design freedom; (3) more efficient risk transfer; (4) longer-term commitment. Through an empirical study of P3 projects, Leiringer was prompted to cast doubt on the correlation between these four mechanisms and actual improvements in innovation. In the light of subsequent developments in research and practice, this skeptical hypothesis was further considered by Gross and Garvin (2011) and by Minchin and Gross (2015). Based on their investigations and their own practice, these later researchers concluded that the evidence strongly favors Leiringer's hypothesis that the innovation drivers traditionally attributed to P3s have not been proven effective and may in fact be largely ineffective. The main problem with any theory claiming that the P3 delivery system encourages innovation is that it must fail to fully appreciate the overpowering imperative of project funding apparatuses (i.e., banks) to protect their investments and to ensure their profits at all costs. The strong profit motive of these lending institutions renders them extremely risk-averse, which runs counter to the necessarily risky nature of innovation. Indeed, their risk aversion may make them innovation-averse.

As for the effect of the P3 model on workplace climate and the relationships between contract parties, the introduction of yet another personality — especially that of a bottom line-focused bank — would seem likely to add more stress and complications to project relationships than would a typical DB project. This proposition is analyzed and discussed in the last section of this chapter.

## Construction-manager-at-risk/construction-manager-as-general-contractor

Like DB, the Construction-Manager-at-Risk (CMAR) delivery system is a fast-track system that permits construction to commence before the plans are completed. CMAR has been very successful in vertical construction, but never caught on in horizontal construction. This lack of enthusiasm in the horizontal construction world is based on the differences between the typical commercial building contractor and the typical heavy/civil contractor. The typical heavy/civil contractor ties up much of the company's net worth in very expensive heavy construction equipment. Because of this large capital outlay, these companies bristle at CMAR provisions that either forbid them to self-perform work or require them to compete with their own subcontractors in sealed-bid processes for any work they want to perform themselves. Seeing that CMAR would probably never become a mainstream delivery system in transportation or other areas of heavy construction, the FHWA turned to Construction-Manager-as-General-Contractor (CM/GC). Though the two systems may differ slightly across agencies, the difference between them that is most salient to this discussion is that CM/GC requires the construction manager to self-perform a minimum percentage of the work tied to the contract, while CMAR either forbids or severely restricts it.

CM/GC, then, is a fast-track construction project delivery system that allows construction to commence on any proposed project site or route, once the property or ROW and required permits have been procured and the design has been sufficiently completed. A major advantage of CMAR and CM/GC over other delivery systems hinges on what constitutes a "sufficiently completed" design. For instance, as discussed above, on a DB project, the owner will generate a conceptual set of plans that potential design-builders will use to generate their proposals. Once a design-builder is awarded the contract, that design-builder will use those plans to generate a final set of plans. In a DBB contract, the owner provides a complete set of biddable plans that potential contractors use to formulate their bids; once the contract has been awarded, the awardee uses these plans to build the facility. If speed is the issue for an owner, a complete set of plans as understood in the DB and DBB contexts is not necessary. On such fast-track projects, since the owner (and/or the owner's representative) and the contractor work with the designer from the beginning, the plans set used by the contractor most often can be seen as "constructable" or much more abbreviated plans than those used traditionally. A relative timeline for a CMAR or CM/GC project is shown in Figure 6.

This abbreviation is possible because, unlike the DB system, CMAR and CM/GC entail the owner's direct contractual relationship with the

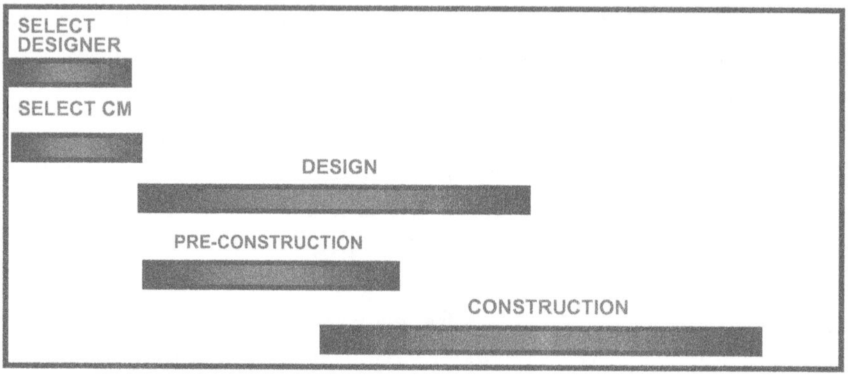


**Figure 6:** A Relative Timeline for CMAR or CM/GC Project.

designer and, hence, the owner's control over the design process. This owner's control over design is a characteristic of DBB that is popular with most owners that use the system. So, CM/GC offers the same advantages as DB (except the single point of contact) along with several advantages not offered by DB such as more speed, a contractual owner/designer relationship, and easier (and probably more) contractor innovation. The greater ease and increase of contractor innovation under these types of contracts is due to the proximity of the parties, the additional collaboration involved, and the freedom from the administrative processes usually required for value engineering under the DB system. Figure 7 shows the organizational chart of a CMAR or CM/GC project.

CM/GC is by far the most flexible of the major delivery systems, and this flexibility gives it its three biggest advantages: (1) innovation, (2) risk allocation, and (3) aggressive project delivery. When it comes to risk, CM/GC and CMAR enable the project team to allocate and re-allocate risks until the party most capable of handling a particular risk at any point in the life of the contract holds that risk at that time. Thus, the risk for a certain item can theoretically change hands multiple times over the course of construction until, in some cases, the risk is eliminated altogether. Conversely, a DB project shifts risk from the owner to the design-builder, but rarely, if ever, in the opposite direction. Researchers now recommend that, in a CM/GC contract, the CM be brought on board at roughly the same time as the designer (Minchin *et al.*, 2014). This simultaneous

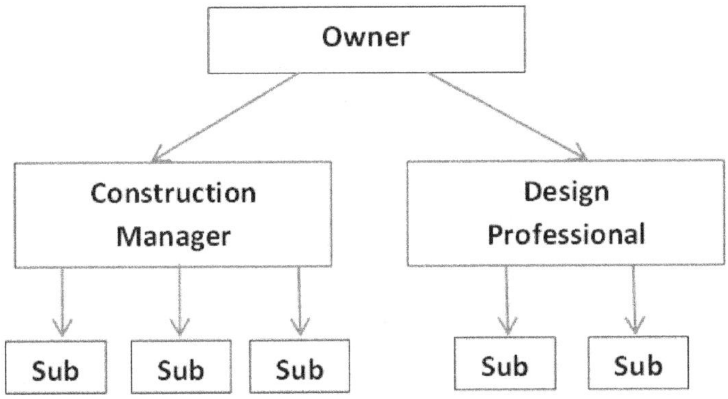


**Figure 7:** Organizational Chart for CMAR or CM/GC Project.

onboarding increases opportunities for innovation, risk allocation and re-allocation, and aggressive project delivery. CMAR and CM/GC allow an owner to get more projects under construction faster than any other delivery method. These capabilities allow CM/GC to be the fastest delivery system from conception to operation of any delivery system (see Figure 7) (FHWA, 2018).

Other advantages of the CM/GC and CMAR delivery systems include consistent cost control through independent cost estimates (ICEs), the ability of owners to hand-pick all team members, a built-in constructability review program (CRP), and the flexibility to change the project scope. Delivery systems that allow for the hand-picking of all team members naturally enable much more collaboration.

## Case study: New construction of a 15-mile freeway through rural and suburban areas

This CM/GC project involved the new construction of a planned 15-mile freeway servicing 13 municipalities. The owner was the DOT of a small state. The project was ultimately extended to 17.5 miles in length, due to $117 million saved by the innovations executed by the contractor. Almost surely, this value could not have been added under any mainstream delivery system currently used in the US, except CM/GC. This project spanned one of the highest concentrations of utility lines in the US, since it was located in a spot where large portions of utility lines connecting the nation's heartland to the west coast breach the Rocky Mountains. Seeing the risks posed by these lines, the contractor had included substantial contingencies in its bid to account for them. Upon receiving the contractor's early post-award cost estimates, the DOT asked the contractor how their prices would be affected were the agency to assume some of the utility risks. Because the reduced risk prices quoted by the contractor were substantially lower than the original quotes, the team was inspired to institute quarterly meetings to identify, assess, and mitigate all the project risks faced by the contractor. Soon, the DOT assumed all risks associated with utilities, and this shift immediately further lowered the contractor's cost estimate. The team then began to go through the same process with the other types of risk.

In the end, the quarterly risk reviews included all appropriate stakeholders, using a process called "decision analysis" to save and set aside about $117 million. The decision analysis process used ranking techniques, risk registers, risk assignment, and risk retirement. Each time a risk was removed or even if risk was lowered, innovation flourished and costs fell. For example, when the utility risk was removed, the contractor devised an innovative solution to span the concentrated utility lines instead of relocating them. The $117 million saved from all the innovation was used to extend the length of the project. Table 2 shows all the

**Table 2:** Details of $117 Million in Savings Due to Innovation and Re-Allocation and Retirement of Risk.

| **Reinvestment** | **Reinvestment amount ($)** |
|---|---|
| Golf Course Reconstruction | $18m |
| Kern River Gas Relocation | $18m |
| Residential Relocation (150 properties) | $40m |
| Kennecott Rail Line Relocation | $11m |
| Rockey Mtn. Power Relocation | $20m |
| Water Tank Relocation | $4m |
| Additional Earthwork | $6m |

instances in which risk was re-allocated to save money. All participants interviewed credited the success of the project to the ease of implementing contractual innovations under CM/GC. This flexibility enabled the highway to be built outside the original contract limits, without a change to the contract documents; thus, a 15.0-mile construction project became a 17.5-mile construction project at no additional cost.

Without the freedom provided by the owner's assumption of its risk, the contractor would never have tried the novel and creative solutions implemented on this project. The success of this and many other CM/GC projects strongly suggests that, because it does the most to foster close working relationships and high levels of collaboration, CM/GC incentivizes innovation more than any other major construction project delivery system used in the US. The next section presents a more detailed discussion of CM/GC's influence on project work environment and collaboration, comparing this influence to that of other delivery systems.

## Comparing the project work environment and level of collaboration within the different project delivery methods

One of the major factors in the quality of life of the people working for any of the parties on a construction project is the choice of delivery systems. In spite of the relatively little available empirical evidence, opinions run strong that some construction project delivery systems foster better working relationships than others, and that, hence, they engender a higher level of collaboration and a more harmonious work environment. Other major influences on project climate and collaboration include project size and complexity and the presence of allowances, or extra funds in some form, in the project budget. If the goal is to have all contract parties work together harmoniously as a team, cushion within the budget is probably the most important factor. From the constructor's perspective, properly funded projects enable the contractor, CM, or design-builder to relax and let go of the fear that they will lose all the possible profit if everything does not go perfectly for them on the project. If the constructor does not have to slash its prices to an extreme in order to get the work, the people working on the project are much more likely to enjoy a highly collaborative and harmonious work environment. This holds true whether the project is low-bid or the owner is immovable in holding an unreasonable position in DB or CMAR negotiations.

The sophisticated owner recognizes that the best relationships are built on trust and will work to truthfully answer all contractor questions — even when it is not necessarily expedient to do so in the short term. Such an owner organization will also work with the contractor to meet all National Environmental Policy Act (NEPA) and environmental permitting challenges. As the cost of a project increases, the pressure seems to rise on the contract parties. This increased stress is probably tied to the fact that higher-priced projects also tend to be the most complex. Also, the fact that larger projects tend to involve more humans and, as the staff size and number of workers increases, so does the probability of interpersonal conflict and the likelihood that someone on the project will either be incompetent or exhibit some other undesirable human trait — as skilled and agreeable as they may be.

Some delivery systems are patently worse than others at promoting a collegial and collaborative work atmosphere. For instance, the DBB system is often singled out and criticized for fostering animosity during a project. A DBB contract almost always requires a construction company to submit the lowest bid for the work — almost always without ever having met the design professional that produced the construction documents. Even though most public agencies and owners of large private facilities arrange a formal "question and answer" session, contractors notoriously loath asking any questions they think might betray their bidding strategy or to share something publicly that they have discovered that their competition may not know. This renders the question & answer (Q&A) sessions less effective, such that many contractors enter DBB projects with significant unanswered questions.

The fact that, in addition to leaving the Q&A session with a lack of critical information, the contractor commencing construction not having met the designer is not ideal. Yet, it is the fact that the contractor organization has cut its profit margin in order to be the lowest bidder that most hurts the relationship between the contractor and the other project parties. In such a situation, a misstep or a delay of any kind could wipe out the contractor organization's profit and might even cause it to lose money on the project. Industry history is full of well-established contractors that went out of business as a result of one disastrous project. Many contractors in such a situation begin the project with a claims mindset, hoping to claw back any profit lost during the competitive procurement process through change orders and supplemental agreements. Other contractors enter the project without such intentions, but once a delay, an accident, rework, or some other misfortune befalls them — whether through their own fault or that of other project parties — the survival instinct kicks in and the company feels that it has to make up the loss any way possible. Once this scramble for survival starts, the relationships between the site workers and their supervisors can deteriorate quickly. Exacerbating this problem are clauses in many contracts (especially public contracts) requiring the contractor to notify the owner of any incident for which the contractor may seek redress within 30 or 45 days after the incident or lose forever the right to seek compensation in time or money. Some owners go even further, requiring the contractor to submit a notice of intent (NOI) to file a claim for any issue for which they might want to seek future compensation. With such an NOI, the contractor effectively retains the right to file a claim in the future for the incident in question, and it must be approved, denied, or rescinded before final payment is made to the contractor at the end of the project. However, as the project progresses and the NOIs start to pile up, the parties involved can grow weary of negotiating the merits of the claims, and feelings and relationships can become frayed over time.

When the DB delivery system first became popular, people immediately noticed that relationships between contract parties were dramatically better than they had been within the DBB system. The terms of a DB contract governing the relationship between the contractor and designer intuitively imply more amicability between them than is possible within the DBB system. Evidence indicates that, for the most part, this is true. Moreover, the rapport between the designer and the owner would intuitively seem likely to suffer for the loss of that relationship. Research also seems to back up this intuition, showing that DPs do not prefer DB for several reasons, but mainly because, after years of currying favor with owners, DB puts the designers on the opposite team. Indeed, they believe that this shift toward an amicable relationship with the contractor on a DB project might erode their relationships with owners and could thus hinder their pursuit of future work.

In spite of the increased amicability between contractors and designers implicit to a DB contract, a recent spike in lawsuits between these DB parties is an ominous development. Participating in a 2019 federal study, one of the top construction litigation attorneys in the nation reported that his firm, which at one point had filed dozens of lawsuits per year against his state's DOT, had only filed two such lawsuits in the preceding two years. He noted that the use of disputes review boards had almost totally eliminated litigation between owners and contractors in his state. He added that the firm's focus is now on representing these same contractors suing designers over design errors or untimely responses while they were partners executing a DB contract (Minchin *et al.*, 2019).

Another negative development related to the application of DB is the recent proliferation of low-bid DB or cases in which best-value DB is arranged so that the financial side of the equation is weighted as equal to, or greater than, the technical component. This formula has left contractors frustrated that they have taken on more risk, with no more reward. This assessment is, unfortunately, accurate. Still, an overwhelming preponderance of evidence shows that, compared to the DBB system, DB improves the project work environment and increases levels of collaboration.

Since DB contracts are at the core of nearly all P3 projects, the project work environment and degree of collaboration are largely, but not exactly, the same on them as on regular DB projects. The P3 addition of powerful new team members such as the concessionaire and lender can easily be seen as likely to thwart some of the advantages offered by DB in these areas. In much the same way that the risk aversion of a lender inhibits innovation, that same caution will almost surely create angst among all P3 parties, since it will likely cause them to feel pressure to avoid mistakes and delays. It will also increase the likelihood of a claim, since the P3 lender stands as another important contract party hypersensitive to any possible cause of project revenue loss.

A P3 project works best when the concessionaire is a faceless entity, void of personality — an outside source acting as a conduit between the owner and the lender and between the owner and the design-builder. This cold and impersonal presence also creates a buffer and an additional layer between the owner and both the contractor and the designer. Typically, the concessionaire knows nothing about, and has no interest in, construction and it is this detached personality (along with the lack of empathy that attends it) that perhaps has the greatest impact on the projects on which it participates. Indeed, such an important yet uncaring contract party is not going to raise the morale of the other project parties.

If administered correctly, the CM/GC construction project delivery system should generate the most collegial work environment and the highest levels of collaboration of any mainstream delivery system in the US. CMAR is identical for the purposes of this discussion, except for one thing, that is, CMAR either completely forbids the CM from self-performing any work or puts onerous restrictions on self-performance of any of the construction. This feature of CMAR could hurt the relationship between a capable and willing CM that desires to self-perform and the subcontractor contractually permitted and hired to do the work. Even more likely to harm the relationship between a would-be self-performing contractor and the subcontractor doing the work would be the scenario in which the contractor tried to procure the work, but has been outbid by a subcontractor for that work. Also, contractors can become aggravated by the time and resources required to bid against its own subcontractors, regardless of who wins the bid and this aggravation can only add to the tension in the relationship between the CM and these subcontractors. Otherwise, whatever holds true of CM/GC is also true of CMAR. CM/GC facilitates a close and collegial relationship among the CM/GC team members (i.e., the parties to the contract) because project cohesion is at the heart of the CM/GC system. For instance, CM/CG calls for all parties to be stationed in close physical proximity to one another, preferably in the same building, so they can meet at a moment's notice to handle any project issues that arise. CM/GC also calls for weekly progress meetings and quarterly ICE and risk assessment meetings. The ICE meetings keep the owner continually updated on the cost of the contract, not only in terms of money already expended but also in terms of any encumbered funds or any costs likely to arise if the current design is pursued. The flexibility offered by CM/GC also enables the parties to assess and re-assess the project risks, and in turn, assign and re-assign risk to the party most able to assume that risk at any given point in the project. This shifting risk burden keeps stress to a minimum and contributes to a collegial work environment, with its resulting high levels of collaboration.

During development of the guaranteed maximum price (GMP), the team members deliberate over why costs may be out of range, asking themselves whether the project is overdesigned, the specifications are overly stringent, the CM or ICE estimates are carrying unnecessary risks, the owner organization can afford the project as designed, and the contingencies and allowances are appropriate. The entire team goes through this deliberative process before moving forward with the design detail, treating each pay item as an individual GMP and agreeing to a reasonable cost for each one. Engaging in this cost assessment very early in the design process enables real-time adjustments to each pay item and to each design detail prior to proceeding to an overall GMP. Costs should thereby remain under control because they are monitored and adjusted during each regular production meeting.

It is no accident that the owner, the CM, the designer, the ICE, and other parties essential to any specific contract are referred to as "the CM/GC team", or simply, "the team". Perhaps, the most important aspect of this team, and the single most important contributor to the collegial work environment of the CM/GC system, is the fact that all team members are handpicked. The ability to choose all their project teammates allows the owner on any CM/GC contract to choose only people who support and believe in the CM/GC system, who share the same work ethic, and who either already do, or likely will, get along well. While no feature of any delivery system does more to position the parties to a construction contract to work together in harmony than does this aspect of CM/GC and CMAR, the one factor that has the greatest effect on the project work environment and collaboration is not unique to any project delivery system, that is, money.

## The most important factor in project harmony and collegiality

Since commerce and travel depended principally on shipping and maritime access until the last half of the nineteenth century, many of the largest metropolitan areas in the US were founded near major bodies of water. While these bays, rivers, intercoastal waterways, lakes, and large creeks are still critical to the commercial engines that power the regional economies that have grown up around them, they often pose significant challenges to those building roadways between them. To help connect adjacent areas of interest over a waterway, modern transportation agencies typically turn to bridges and tunnels. The recent completion of one such project involved the improvement of two existing two-lane tunnels, built beneath a river in 1952 and 1962, respectively. Connecting two cities, one of the tunnels went into and out of a downtown area and one went into and out of a midtown area. The downtown tunnel had been expanded to four lanes in 1987, but the midtown tunnel remained in its original configuration and had become the busiest two-lane road in the state — carrying over 35,000 vehicles per day. Soon after passing the Public-Private Transportation Act in 1995, the state legislature received several unsolicited proposals to add two more lanes to the midtown tunnel, upgrade the downtown tunnel, and construct a freeway extension linking the two. Approval from the city councils of the two cities involved was necessary before evaluation could proceed, but one of the city councils voted against the proposed tolls in 1999, scuttling the effort. The state DOT revived the project in 2004, requesting expressions of interest from the private sector and then soliciting conceptual proposals in 2008. Only one firm submitted a proposal for the work.

Although an interim agreement was signed with the sole proposer in 2010, this P3 project grappled with the challenge of developing a contract approach that satisfied the stated pricing goals. The local politicians wanted a toll of $1.50, even though the private sector proposal for the project estimated initial tunnel tolls between $2.00 and $3.00. This discrepancy complicated negotiations for this DBFOM contract, but the issue quickly came to a head when the mayor of one of the cities published an open letter to the governor in a local newspaper, demanding the $1.50 toll. Despite even more contentious negotiations, played out largely in the press, all parties finally signed a $2.1 billion P3 contract. Although the project was built and now functions very well, the general public almost uniformly despises it because, contrary to early promises, the $1.50 tolls had to be raised. However, the ongoing political contention and media controversy surrounding this project stand in sharp contrast to the collegiality and congeniality of the parties to its P3 contract.

Observers in fact noted that never did a major project have a more congenial worksite nor has one enjoyed a higher level of stress-free collaboration. This harmonious work environment was in no small part due to the fact that, because the project had only one proposer, the owner paid top dollar for the work. Estimates by state officials put the profit made by the design-builder on this contract at $200 million. A principal in one of the main parties of the contract admitted that this project had netted half of his large company's profit for the fiscal year. Such statements indicate that the project did indeed have plenty of resources, which, again, no doubt contributed significantly to a remarkably harmonious workplace. Despite the large size of this project, its delivery by a system not particularly noted for its ability to foster collegiality, and surrounded by controversy, the project generated no lawsuits, no claims, no supplemental agreements, and fewer change orders (six traditional change orders, plus five no-cost administrative change orders, plus a number of routine fuel-price adjustments whose cost was incorporated into the contract via change order) than could have possibly been anticipated (Anonymous, 2020).

This outcome supports the theory proposed at the beginning of this chapter, that a sufficient degree of comfort felt by the private contract parties (especially the contractor) — bolstered by their awareness that the contract provides for more than enough money to protect against small mistakes and minor misfortunes — makes for the most harmonious working relationships and, thus, the highest levels of collaboration.

For projects not so awash in cash, however, the choice of delivery system probably plays the biggest role in promoting the professional accord of the contract parties and the desirable degree of collaboration resulting from their harmonious relationships. The introduction of integrated project delivery systems to the industry has profoundly affected the way contractors, owners, and designers relate to one another on construction projects in the US. So much, in fact, that one respected researcher claims that the proliferation of DB, CM/GC, and CMAR has caused blue- and white-collar construction personnel to become accustomed to treating each other with more respect and courtesy than in the years when all construction was accomplished under the traditional DBB system. Moreover, that investigator notes that this cultural shift toward a more congenial work environment has even spilled over onto DBB projects (Gransberg, 2017).

## Conclusions

Of the many conclusions that can be drawn from this discussion, the following three stand out: (1) different project delivery systems do create different work environments and levels of collaboration; (2) integrated project delivery systems (as well as other Alternative Contracting Methods (ACMs)) foster a more collegial work environment and higher levels of collaboration than does DBB, with CM/GC offering the greatest opportunity for such outcomes, followed by CMAR and DB; (3) a project with a budget sufficient to allow for small mistakes by the contract parties eliminates the contractor's fear that any delay or mistake will wipe out all its profits, or worse, drive it out of business. This does more to create a positive work environment and a higher level of collaboration than any delivery system can do.

Lesser conclusions include the observation by the author and others that the use of integrated project delivery systems has led to more congenial relationships on projects across the board, even on DBB projects. As people get accustomed to the idea fostered by ACMs, that the other party is not the enemy, that attitude is slowly finding its way into the "hard dollar" DBB contracts. Also, in states where ACMs are commonly used in conjunction with disputes review boards, contractor claims against the owner have almost disappeared.

Several research threads supporting these conclusions could be further developed to contribute to the collective body of knowledge on construction project delivery, procurement, and contracting. Besides contributing to the scientific heft of these accumulated findings, the conclusions shared in this chapter have the added advantage of ringing true intuitively in every respect. However, additional research of a more quantitative nature in this area would be useful. Also, were a researcher be able to procure accurate data on contractor profit margins, a qualitative investigation into the relationship between these margins and project work environment/collaboration among the contract parties would be useful. Lastly, research into any ties between project work environment/collaboration among contract parties and measurable positive project outcomes would be valuable. For example, to establish a correlation between a positive project work environment and project cost overruns would be a worthy goal, as would establishing a connection between a high level of collaboration among contract parties and construction project time extensions. Tying project work environment and level of collaboration among contract parties to construction quality would also be of great value. Research into these areas would be most welcome as reliable accounts of the ongoing history of construction as well as a tool for predicting the success of individual projects and future trends in construction project management.

# About the Author

**Edward Minchin** is a Professor at the Rinker School of Construction Management at the University of Florida. In his 35-year career, Dr. Minchin has worked as part of eight areas of the construction process — laborer, contractor, designer, inspector, owner, researcher, consultant, and adjudicator. He has 145 publications, including two books and seven book chapters. More of his publications are in the area of project delivery than any other topic. Dr. Minchin currently simultaneously holds three professorships: the UF Term, UF Research Foundation, and Holland Professorships. He was an invited keynote speaker at four international Construction Conferences in Panama, China, Canada, and Thailand.

# Chapter 4

# Early Contractor Involvement in Project Delivery — Drivers and Impediments

Ghada M. Gad[1] and Edward Minchin[2]

[1]*California State Polytechnic University, Pomona.*
[2]*University of Florida*

## Abstract

Owners and contractors have long encountered a deep disconnect at design hand-off, when the project structure proves to be unbuildable as set out in the plans. This gulf between the owner and the contractor is especially common in traditional project delivery methods. Indeed, contractors are constantly heard saying, *I would have completed the job for half the price in less time if I had been able to provide earlier input into the design.* Owners who have been able to involve contractors early attest to the added value accrued from their construction knowledge and experience and designers developing project plans in silos simply cannot replicate this invaluable contractor insight. Although the industry has long faced this issue, owners still have trouble getting contractor input earlier in the process, given existing legal and/or cultural constraints. However, some owners have successfully operated within these constraints to capture the added value of a contractor's input in their designs. Early contractor input increases innovation, makes project plans more constructable, and expedites the entire process. This chapter examines the drivers of early contractor involvement in the light of the many factors impeding it. The authors review and describe the effective methods used by owners to allow contractor design input at various project stages (from planning to construction) and under different project delivery methods. These successful owners have used the standard Contractor's Q&A method during bidding and the Value Engineering Services method as well as more innovative ones such as Alternative Technical Concepts (ATCs) and online input solicitation platforms. A case study also demonstrates one owner's successful use of a method that solicits pre-bid contractor input during contract document development to enhance the project's constructability.

## Introduction

Ever since people started using sets of plans to build structures, miscommunication between designers and constructors has bedeviled project construction. The advent of the Age of Specialization brought about the separation of the designer from the builder, a shift that naturally increased the likelihood of builders not understanding the intent of designers. Specialization meant that the engineers designing a project would have nothing to do with building it. Soon, specializations formed tiers, with new specialties emerging on the design side, e.g., structural, geotechnical, and hydraulic engineering. Contractors also splintered into new specializations, e.g., structures, earthwork, or paving. The days of the "generalist" were numbered.

The traditional Design–Bid–Build (DBB) delivery method that came to dominate the American construction industry is a "linear" system characterized by a gap between the completion of the design and construction documents and the moment the job is let. The time between design completion and project letting could be very lengthy with the design professional and the constructor sometimes never even meeting. As DBB became more prominent, the walls of the "silos" within each of these project personnel groups got higher, and the higher the wall rose, the more problematic the miscommunication became. As life sped up, the linear system began to lose favor, and the industry began searching for ways to improve designer–contractor communication and overall project performance. One of these approaches to better communication, now called

Early Contractor Involvement (ECI), aimed to elicit contractor design input at an earlier project stage before commencement of construction. [Note: In the section below, ECI is not discussed as a project delivery method (PDM), but rather as a set of strategies used by owners to involve contractors at earlier project stages].

## What is early contractor involvement?

The main objective of ECI is to involve the actual constructors of a project at an early enough stage to ensure that their recommendations will have the least negative impact on cost and schedule. Their recommendations are solicited to improve the "constructability" of the project, i.e., the level of ease and efficiency with which the project can be built from the construction documents provided. The term "constructability" was coined at the 1983 founding of the Construction Industry Institute (CII), when it was set as an initial research target. The major tool developed by researchers in this domain was the constructability review, which in turn engendered the constructability review process (CRP) that now extends beyond the pre-construction phase. The added value of constructability is realized through project outcomes such as reduced construction costs, shorter project durations, and enhanced maintainability (Conner and Tucker 1986). The CRP came out of the battery of early CII research into constructability and gained great traction from widespread industry use in recent decades.

CRP is not the only way to achieve ECI; the industry has implemented a number of quite innovative techniques, all with the goal of designing more buildable projects. Some of these techniques include a contracting method that sets contractual terms enabling project team members to work together closely. These ECI-friendly contracting methods are built into alternative PDMs such as Design–Build (DB) and Construction Manager/General Contractor (CM/GC). Other approaches to ensuring ECI include the solicitation of innovative ideas for improving the baseline concept and design, whether pre-award (through Alternative Technical Concepts) or post-award (through required Value Engineering or Constructability Review Services). Some owners have gotten very creative, soliciting contractor feedback by using the contractor organization as a third-party consultant before the release of project bid documents.

However, as environmental regulation has become more stringent, the high level of constructability enabled by the research of the late 1980s and early 1990s has been eroded by two major trends: 1) roadway and accompanying drainage designs that previously had to be approved and permitted only by local/state water management districts have come to require many more permits from other affected federal environmental agencies, e.g., the US Coast Guard, the US Army Corps of Engineers, and the US Fish and Wildlife Service (NCHRP, 2019); 2) once permitted at the local and state levels, project designs have required increasing scrutiny in a process stipulated in the 1970 law known as NEPA, the National Environmental Policy Act.

Though NEPA has been around for decades, its enforcement has gradually become more stringent, and agencies now consider it a major milestone in the life of any highway project. The collision of alternative PDMs with these expanded and intensified environmental-permitting processes has challenged owners across the country in their efforts to implement ECI effectively. Far too often now, contractors submitting excellent ideas for improving quality, lowering costs, or shortening durations of the design construction process of highway and bridge projects have these ideas rejected because their implementation will trigger a NEPA re-evaluation or the re-execution of an environmental-permitting process. To assure the many advantages of ECI, researchers are increasingly encouraging owners to procure their contractor or design-builder in the early project phases, concomitant with the start of the NEPA and environmental permitting processes.

## Why ECI?

ECI is widely understood to be essential to elevating the quality of project design and construction. Indeed, public agencies and large private construction owners that have practiced early contractor involvement have overwhelmingly testified to its merits. Contractors that have been brought in early have unanimously voiced their preference for this timely involvement. The intent of ECI is to identify ambiguous specifications, refine project design, increase project construction efficiency, forestall potential disputes, reduce cost overruns, and prevent schedule delays. One of the major reasons for ECI is constructability. Research has documented that conducting Constructability Reviews generates key benefits that extend past a project's pre-construction phase. These benefits include reduced costs, shorter schedules, improved quality, better risk management/control, fewer claims and change orders, and enhanced maintainability.

Many studies have assessed the benefits of ECI, beginning with the O'Connor and Tucker (1986) study that determined six types of constructability improvement types present throughout the project development lifecycle: (1) construction-sensitive design; (2) effective communication of engineering information; (3) innovative construction techniques; (4) efficient construction resource management policies/standards; (5) vendor/subcontractor improvements; (6) construction input into the design. Ford *et al.* (2004) researched the main benefits of CRs on construction projects (with a focus on reducing duration) and showed that increased CR efforts ensure a shorter construction phase, even if they cause a slight increase in design phase duration. Furthermore, implementing a formal CR process ensures a reduction in the time required per review, as compared with traditional one-off CRs (e.g., 1 hour versus 2.5 hours). This efficiency underscores the need for all owners to adopt an integrated CR program.

In the industrial construction sector, constructability has been called the optimum tool for reducing the facility investment risks associated with advanced technology facilities. Such facilities are usually delivered through fast-track PDMs that are similar to DB. Any study of the manufacturing and construction industries will inevitably involve Total Quality Management (TQM) and VE. When constructability is considered in this context, three elements emerge as integral to obtaining the requisite quality — whether of a manufactured good or a construction project: TQM, VE, and constructability. These three elements focus on different aspects of the production process. TQM focuses on the emerging quality of the manufactured or constructed product, VE concerns itself with process functionality, and constructability focuses on improving the buildability of the product (Chasey and Schexnayder, 2000).

Advances in many technologies and tools — particularly 3D modeling (Chong *et al.*, 2014), automated review models (Jiang and Leitch, 2016), and GIS-based systems — have enabled ECI to provide many projects benefits. As these and other applications continue to evolve, these benefits will only increase. For instance, a BIM-based project with 4D construction simulation can generate a structural simulation of the facility's design, which designers can then utilize to determine its constructability. In addition, by linking the BIM model with constructability factors (e.g., resource availability, site impacts, and weather effects), project teams can deduce a constructability score to gauge the constructability of the design (Chong *et al.*, 2014). A GIS-based system can also improve the constructability of a design by inter-connecting its different features and by interfacing with other existing models and databases to form the completed design elements. This full picture provides the reviewers with all the information they need during the preconstruction review, prior to calling final designs for bid.

## What are the different forms of ECI throughout the project phases and within the different project delivery methods?

Owners have tried many methods throughout all project phases to simplify and standardize the integration of contractors' design ideas into contract documents, applying some during planning and prior to the procurement stages (e.g., Value Engineering Proposals and Alternative Technical Concepts) and applying others post-award (depending on the PDM used or the CRs conducted). Table 1 presents the methods most widely used by owners. Even though bidders' meetings and Q&As could

**Table 1:** Different ECI strategies throughout project phases.

| | Phase | | | |
|---|---|---|---|---|
| **Involvement of contractor through…** | **Planning** | **Prior to letting** | **During procurement** | **After award** |
| Value Engineering (VE) | x | x | x | x |
| Owner Portals | x | x | | |
| Alternative Technical Concepts (ATCs) | | | x | |
| Project Delivery Method (PDM) use | | | | x |
| Constructability Reviews (CR) | x | x | x | x |

be considered means by which owners receive contractor feedback prior to award, they are mostly driven by competition and thus do not offer a platform for open feedback and communication.

### *Value engineering (before or after award)*

VE is a standard practice in many construction projects that aims to "improve the value of a project without compromising its functions" (Mao *et al.*, 2009). It ensures the systematic consideration of any alternatives that would provide the same function more efficiently and with lower total overall costs (i.e., construction, operation, and maintenance). VE can be done at an earlier project stage, once the design is complete or nearly complete. It can be conducted by a designated team within the organization or by external consultants hired to review the design. It can also be folded into the preconstruction services contractually required of the contractor under a number of PDMs. For example, under DBB, the contractor's scope could include the performance of VE to provide alternative solutions for reducing project costs. While VE focuses on the overall reduction of lifecycle costs, constructability concentrates on optimizing the construction process itself. Thus, constructability and VE are not mutually exclusive, but rather can be implemented so as to complement one another (Russel *et al.*, 1994).

### *Soliciting feedback from contractors prior to release of bid documents*

This type of "industry review" is considered innovative in that contractors are given a project's plans and specifications and asked to provide feedback and comments prior to the release of the bid documents. As detailed in the case study below, this innovative approach was successfully recently used by the Utah Department of Transportation (UDOT) to solicit input from contractors on specific projects.

### *Alternative technical concepts*

The FHWA Every Day Counts (EDC) project has brought national visibility to the implementation of ATCs on transportation projects, defining an

ATC as "a request by a proposer to modify a contract requirement, specifically for that proposer's use in gaining competitive benefit during the bidding or proposal process…[and] must provide a solution that is equal to or better than the owner's base design requirements in the invitation for bid (IFB for DBB) or request for proposal (RFP for DB) document" (FHWA EDC-2). ATCs have added the completely new parameters of creativity and innovation to bidder selection criteria. They allow the owner to review contractor design input in writing prior to contract award and then assess its possible benefits in terms of increased constructability, cost savings, and schedule reduction (Carpenter, 2012; Coblentz, 2012; Horn, 2012). Another major advantage to having such well-documented contractor input so far in advance is the ability to identify issues within the solicitation before contract award and before such issues lead to disputes (Gad *et al.*, 2020).

ATCs have been implemented in virtually all PDMs, but they seem to have become an inherent part of the CM/GC and the Progressive DB (PDB) delivery methods. For instance, for DBB projects, ATCs are allowed to be incorporated during procurement (after design), while for DB and CM/GC projects, they are included in the RFPs. The only difference between DB and CM/GC is that CM/GC allows for ATCs to be incorporated even before a design contract is awarded, once the pre-construction service contract is awarded. CM/GC and PDB both facilitate ATC implementation and eliminate the need to extend the design baseline to incorporate ATCs (Gad *et al.*, 2015, Gransberg, 2014).

To realize the benefits of ATCs, projects must have well-structured implementation mechanisms (Gad *et al.*, 2014). To determine which proposed ATC to approve for a project, owners are certain to conduct preproposal confidential one-on-one meetings with the proposers. Moreover, they understand the importance of not restricting ATCs from non-ATC sections of the bid document, as long as they do not conflict with the NEPA. The contractor assumes responsibility for the environmental and schedule impacts of such ATCs and must prepare the documentation submitted for approval. Essentially, the contractor is the party that decides whether or not to take the risk. If the contractor organization decides to submit the idea in its proposal, the owner's ATC review team conditionally accepts it (with the condition of getting the permits) and explicitly states whether it will cause any environmental impacts. When a stipend is offered for each proposed ATC, the owner organization can gain the additional value of using the ideas submitted by the non-winning bidders and direct the winning design-builder to use any of those bidders' ATCs.

## *By virtue of the project delivery method*

Some PDMs, such as CM/GC and DB, by definition institutionalize the constructability process. For example, CM/GC has early and consistent contractor involvement woven into its processes as does no other system. In CM/GC, neither a value engineering nor a constructability component is necessary because the proper execution of the CM/GC process entails both of these concepts as a matter of course. A CM/GC hired at 10% design completion is tasked with preconstruction services, one of the most important of which is constructability review. The DB PDM makes it easier to exploit the contractor's knowledge in the design phase than under the traditional DBB approach; a single DB team composed of both designer and contractor work together to develop the design.

## *Constructability reviews (after award)*

An important distinction between a CR and a Plans Review (PR) merits attention before further discussion of CRs. The PR is a traditional industry process comprising design quality control and quality assurance (QC/QA) reviews of the design plans and other project design documents during pre-construction. PRs typically include the determination of whether the designs meet the required industry functional/technical standards for the work, and whether the designs are economical, constructible, maintainable, and appropriate for the specific project conditions. PRs typically involve checklists that address format, compliance, consistency, constructability, calculations and data, and other aspects of the plans that ensure compliance with the organization's requirements (e.g., processes, procedures, standards). In short, PR is primarily focused on activities internal to the organization. Notably, while the performance of a PR entails the need for plans by definition, a CR can be conducted entirely without plans.

In the design phase of a construction project, design firms pursue certain constructability efforts, such as determining the preferred timing for constructability implementation, exploring the factors that promote or constrain constructability, and determining the benefits of a formal CR program that accrue to the designer. Among design firms, "peer review" and "feedback systems" have proven to be the most popular tools used to conduct CRs. The successful implementation of a continuous CR process entails four main elements:

- Institutionalization of the CR process champion.
- Emphasis on a quality-driven rather than a schedule-driven design process.
- Clear, yet flexible guidelines for executing CRs across the entire range of project types and sizes (CRs at specific points of PDP).
- Involvement of construction contractors.

Over time, the many benefits of the CR process have been hampered by the increasingly stringent NEPA and environmental-permitting processes. Of these three typically scheduled and coordinated processes, the NEPA and environmental-permitting processes are often completed before the CRP commences. This sequence forces owners to reject numerous ideas from contractors — ideas that would improve quality, reduce cost, or reduce project duration — because implementing these worthy ideas would trigger a NEPA re-evaluation and/or a replication of the environmental-permitting process.

## What are the barriers and drivers for success of ECI?

Figure 1 summarizes the various barriers to and drivers of ECI success that will be discussed below.

### *Barriers and impediments to ECI*

Many would say that the source of most of the barriers to ECI is the segregation between design and construction. These barriers principally

**Figure 1:** ECI barriers and success drivers.

involve poor communication practices. Ironically, nothing could replace segregation with integration as effectively as would ECI. The many barriers to successful constructability implementation cited by researchers and industry professionals fall into the following categories: (1) organizational, (2) staff, (3) legal, or (4) cultural.

*Organizational barriers* include the absence of feedback systems, poor team organization, lack of communication, and insufficient time for reviews. The lack of a standard and formal constructability system results not only in lower quality construction on any particular project but also in a paucity of construction knowledge input into that project's design phase and that of all future projects (Jergeas and Put, 2001). Having substandard lessons-learned programs exacerbates the serious problem of inadequate communication tools, which hinders the appropriate use and transfer of constructability knowledge. Focusing on duration reduction benefits, Ford *et al.* (2004) found that for any CR program to be successful, the actual implementation of CRs should not exceed any project duration savings they garner. Lack of communication between designers and constructors, low-quality plans and specifications, and insufficient time for reviews have been identified as the most critical issues hindering transportation agencies from fully benefiting from CRs.

*Staff barriers* include the designers' lack of construction knowledge, inadequate agency training, a shortage of qualified personnel, and ignorance of constructability benefits (Kuo and Wium, 2014).

*Legal and contractual barriers* include any issues associated with an agency's limited available procurement methods and the way the procurement method is set up. Moreover, contractual clauses that establish or reflect an untrusting relationship often impede the open communication and collaboration needed for successful CR.

*Cultural barriers* include any issues associated with an agency's organizational culture. One of the most difficult questions an agency can ask is whether it is prepared to adopt and implement an innovative practice. Answering such a question honestly is challenging because it forces the agency to come to terms with its organizational culture. Because the decade-long use of the inherently silo-making DBB method has so fundamentally shaped employee perceptions and organizational structures and practices, implementing any approaches that allow or even require ECI constitutes a major paradigm shift for a state agency. Early studies found that "as agencies attempt DB for the first time, they are constrained by the low-bid culture in their organizations" (Molenaar and Gransberg, 2001). Of course, most low-bid contracts eliminate ECI. In addition to cultural barriers to ECI within organizations, such cultural barriers can also stand between organizations (Chasey and Schexnayder, 2000). For example, trust issues with the contractors, reluctance by owners and/or design professionals to share information, or lack of proper knowledge within the local contracting community all create barriers to proper communication and openness to new methods.

## *Recipes for ECI success*

To overcome the different types of barriers and optimize ECI quality, benchmarking constructability implementation is critical during the project development process (Hiley and Yagci, 2001). Moreover, most researchers and practitioners agree that constructability should be applied early in the project development process to overcome these barriers and to realize the desired benefits (with consideration of the required tools and technology) (Khan, 2015). Another way to remove barriers to constructability is to establish ongoing education and training for continuous improvement and to take a multidisciplinary team approach to projects (Hiley and Yagci, 2001). An effective lessons-learned program is one of the best ways to ensure the long-term benefits of constructability as is the use of a feedback system to transfer construction knowledge (Kuo and Wium, 2014). Owners have long recognized the need to implement constructability, but have found that the resource- and labor-intensive nature of the CRP has limited their ability to formally implement it. In addition, few applications properly evaluate the cost–benefits (C/B) of CR implementation. However, for a CR program to be effective, any cost or schedule savings in the construction phase must be greater than any added costs or time needed for conducting the reviews during the design and planning phases. Thus, resources must be allocated properly for optimum constructability management.

For effective implementation, agencies must determine which changes are necessary to establishing new working relationships with contractors, suppliers, and consultants. Such challenges to changing the project delivery approach are common when an agency adopts any innovative delivery method (e.g., DB or CM/GC) and successful change often depends on an agency's formal and informal cultures. When an agency is procedurally rooted in traditional means and methods, it is likely to face varying degrees of resistance to innovative ideas and approaches. Instead, the agency's *formal culture* should be open to innovation, risk-taking, and improvement of the *status quo*. Similarly, the agency's *informal culture* must support any new process that it wants to succeed fully. Informal culture consists of the ways an agency actually gets work done, apart from procedures and policies. It is often a natural response to gaps in these official guidelines and is usually a result of the need to adapt the rules to change. Agencies should never underestimate the impact — positive or negative — of the informal culture on the implementation of any innovation, including ECI. Within agencies' informal organizational cultures, any initial negative perceptions of an innovation have been found to slow down or stop its use. Sometimes, though, the initial lack of success of an innovation, or even its failure, was caused by specific issues (e.g., statutory exclusions of procurement methods other than low-bid or lack of proper knowledge within the local contracting community). After having experienced and resolved these issues, agencies have learned to overcome initial informal cultural resistance and successfully restart the implementation processes for innovative practices of all sorts.

The culture — both formal and informal — of any organization begins and ends with its personnel. Successful implementation of innovative methods requires the support of in-house personnel and staff. Several staffing issues may affect the success any innovation and can be grouped into the following five categories:

1. *Personnel assignments*: Project personnel must be innovative and open to new ideas in order to advance the application of any innovative methodology. If an owner organization wants to implement ECI, for example, it must hire individuals enthusiastic about such programs, rather than people who will likely resist participation even when assigned to administer them. Often, to select the right personnel in time for the beginning of construction, the management team will meet at the outset of the project to determine the level of expertise needed. In addition to assessing expertise, the level of enthusiasm for ECI must be determined in the same way.
2. *Staff availability*: Projects that implement ECI tend to be integrated, fast-track projects. They, thus, require more of a time commitment from staff, particularly for the design review process. For instance, as a design-builder's design production rate rises and falls, the agency's design review load fluctuates with it. Indeed, if the contractual agreement does not regulate the review flow, an agency may be forced into accommodating the design-builder's pace.
3. *Staff capability*: DB projects need personnel who can manage the issues and complexities unique to the DB method. The same holds true for CM/GC. In order to succeed in their respective roles under any fast-track system, staff members should be able to work under pressure, demonstrate flexibility, and handle multitasking as needed. Because these schedules are accelerated, the staff also often needs to be knowledgeable in project control practices. Since agency staff size is often smaller on integrated projects, individuals with high levels of expertise in their respective technical areas may be required to perform independent design reviews or oversee consultants involved in the review process. Agencies also often create permanent units

dedicated to supporting the implementation of targeted innovative practices across project portfolios. Similarly, such support units can be developed for individual projects. Such units are ideally staffed with agency personnel experienced in whichever innovation is being implemented. Failing that, the units should be staffed with people enthusiastic about the innovation and excited at the prospect of being part of the success achieved through the innovation. Even in the absence of any such support unit, individuals with experience are often used on an as-needed basis to help make any innovation implementation a success.

4. *Training*: Staff may need additional training during the design phase to understand and perform the duties required of owners executing projects with new methods. For instance, ECI presents challenges of proprietorship, anonymity, and secrecy that are not even related to the obvious problems of resistance to change, jurisdictional disputes, and cultural change inherent to the implementation of any innovation. Promptly assessing these needs and then training these individuals accordingly are crucial to a successful innovation implementation. For example, if an agency is allowing submission of ATCs for the first time, staff must be quickly trained on the confidentiality issues that attend this approach to design management during procurement.
5. *Utilization of consultants*: If an agency does not have the required in-house resources to develop and/or manage ECI, it may need to consider hiring outside consultants. The agency should address the following questions to determine which roles should be filled by consultants:

    (1) Will the general engineering consultant role be outsourced to a private engineering company to manage the implementation of this particular innovation?
    (2) Are the tasks for the consultants all traditional enough that sufficient contract language exists or will a new language need to be developed?
    (3) Which roles will these consultants perform in defining the methodology used on a project?
    (4) Will the consultants use a standard already in place or is the owner/agency expecting them to deliver/develop a new standard?

(5) With respect to roles and responsibilities, is the agency willing to accept the major changes inherent to adopting an innovative delivery method?

## A Utah DOT case study: Thinking outside the box

The Utah Department of Transportation (UDOT) is a decentralized agency staffed by around 1800 employees, with four regional offices statewide. These offices are largely autonomous in terms of their approaches to management, including the ways they solicit contractor input. A discussion of UDOT's planning and development strategies for construction projects will provide context for the various ways UDOT has innovatively involved contractors at early project stages. In the planning process, projects are green-lighted and then assigned to the different project managers responsible for assembling the teams to design and deliver the project. Designers are usually outsourced experts in various design areas, with 85% of design performed by consultants and very little developed in-house. Each regional office has around five to seven project managers who run their projects from cradle to grave. UDOT has been unique in its record of quick project turnaround — with scope defined and funds available, they can bid out a project in less than a year. Being a decentralized DOT, UDOT does not require its regional offices to perform a formal CRP nor are individuals within the same region required to follow the same formal process. This flexibility enables its personnel to be innovative in the ways in which they can involve contractors early in the process. The section below discusses some of the many successful methods UDOT has used to solicit early contractor input, specifically online platforms and industry outreach efforts.

### *Platforms to solicit ECI*

*"Project Preview Website" to accommodate comments prior to project advertisement*: On traditional projects, the first time a contractor sees the project design is typically at advertisement. However, a few years ago, UDOT started inviting contractors to preview projects (especially larger ones) online prior to advertisement. This platform is believed to have

saved UDOT millions of dollars, since it has given the contracting community a mechanism for providing comments on specific project plans and specifications before advertising. It allows contractors to anonymously post their comments and questions on sets of plans set to be bid out, but prior to bid release. These questions and comments are screened and responded to within 48 hours. From an owner's perspective, this approach is invaluable, since it provides an unbiased platform for contractors to speak freely and frankly, enabling an openness that is often absent in conventional face-to-face conversations with prospective builders. Comments are reviewed thoroughly, since none of the recommendations/questions anonymously provided by the contractors should provide an advantage to one over the other once the bid is out. The totality of the contractor feedback is considered, and changes are made on the project bid documents accordingly (UDOT, 2019a).

The Project Review Website has received a lot of positive feedback from the contracting community, since it has given contractors a means of forestalling problems they clearly see coming in the pre-release bid documents. In some instances, UDOT received up to 96 RFIs from contractors using the website, a response that prompted the agency to issue two addenda. UDOT could further analyze the data received on this platform and use them to formulate a metric for constructability reviews. Although this website clearly offers many advantages, some contractors have voiced confidentiality concerns, worrying that, if they were awarded the project, their comments might be used against them as a liability issue were they to request any change orders during construction (UDOT, 2019b).

UDOT has utilized the project preview website in some interesting and impactful ways. On its Mid-Valley Highway Project, for example, two weeks prior to bid, the agency invited seven contractors — all of whom had provided feedback on the website — to a follow-up preview meeting for further discussion of their comments, specifically those related to geotechnical issues. Most of their comments were related to measurements and payment items. According to UDOT, such meetings are much more effective than traditional pre-bid meetings, which can be replaced simply by means of a waiver letter signed by the contractor organization confirming its understanding of the bid requirements. After the bid goes out, contractors no longer share much information, since the process then becomes competitive; thus, the pre-bid meetings offer a great venue for open project-specific discussions. UDOT also sought feedback from contractors on certain risky items, one of which was a shipment of a 220-foot-long girder bridge needed for a project. On another DBB project, the regional office sought contractor input to get a sense of the prospective bidders' approaches and to improve the bid document. Specifically, the idea on this project was that the region was trying to determine whether ATCs would work on DBB projects. For the I-215 project — a relatively complex project with a \$100M construction budget — UDOT sought specific answers to a list of 20 questions. Four to five contractors responded to at least some of these questions.

In spite of its tremendous value to UDOT, one drawback of this approach is that it does not require contractors to commit to answering the questions and providing feedback as would be required from a contractor hired for this purpose (as explained below in the "out-of-state contractors" section). Moreover, from the contractor perspective, the time-consuming nature of the preview/input process poses a degree of risk, since, even though these previews can help the contractors as well as UDOT, the busy market sometimes does not permit them much time for providing input on projects they ultimately may not even bid on. The most effective input that could be provided would relate to maintenance of traffic and constraints on limits of operations (e.g., lane closures, timings, barrier use, and daytime or nighttime operation).

*Out-of-state contractors*: The practice of hiring an out-of-state contractor for early project input was used by UDOT on one of their DB projects (I-15 South Project) to address its concerns about the project's constructability and to solicit advice on handling various unique construction issues (including collectors and ramps positioned around a railroad, with a bridge being cut longitudinally). To ensure the constructability of this one-of-a-kind project, UDOT solicited an out-of-state contractor (one ineligible to bid on the project as a contractor or as a subcontractor) to review, critique, and report on its concept plans. This contractor was hired by the program management team as a sub-consultant for the sole purpose of performing a constructability review. The contractor organization was not given any specific protocol to follow; it simply offered its assessment of what would happen during construction, providing truly valuable insights. The organization had to sign conflict of interest and confidentiality forms.

*Alternative Construction Request (ACR)*: ACRs are based on the same concept as ATCs, with a key difference being that contractors are expected only to suggest modifications to means and methods rather than actually providing alternative technical designs. This mechanism got varying reactions from the four regional offices, with only one having successfully adopted it on their projects. In the DBB context in which ACRs are used, the contractor organization was expected to submit its ACR within the six-week advertising period. In this process, after award and before NTP, the specifications are revised to reflect the ACR.

*Staff augmentation program*: UDOT has yielded great benefit from hiring consultants whose key service is staff augmentation. Because most of these consultants have worked as contractors for 30+ years, winning and losing bids, they bring an experienced contractor's perspective to their UDOT staff positions. They serve as project advisers by conducting constructability and estimating reviews. They are hired for a three-year term based on their responses to Central Office (CO) RFQs, yet they serve the various regional offices as well as the CO. Thus, they are co-located in the different regions, according to project load by region.

## *Committees with contractor representation*

*Standards committee*: UDOT has assembled a standards committee with contractor representation, and when new standards are proposed, the contractor members contribute their feedback to the review. Meeting once every other month, this committee provides a forum for reviewing any proposed changes. The feedback on these changes in standards is passed onto UDOT for acceptance or rejection.

*American General Contractors (AGC) committees*: UDOT adopted an idea implemented by the Washington DOT, which had AGC subcommittees provide a platform for contractors to directly communicate with the agency. UDOT sends design plans to its AGC committees, seeking constructability input from the contractors. One of these AGC committees is the Heavy Highway Committee, which has executive-level representation for input on general issues applicable to all projects. Being a relatively small DOT, UDOT does not have the staff resources for the amount of project review it needs, so the unique working relationship UDOT maintains with AGC member organizations promotes open communication and information gathering.

*Contractor summit*: UDOT holds a yearly summit to provide a venue for preconstruction engineers to communicate with contractors about the general issues they face during project construction.

UDOT is recognized nationally for its innovative approaches to project delivery, a characteristic that UDOT personnel generally attribute to its smaller size. Its small size and decentralized organization allow personnel and contractors alike to implement and be rewarded for their innovative ideas. In spite of the many benefits of this relaxed approach, careful management is crucial, since too much flexibility could be a disadvantage if personnel lack experience and training.

## Concluding remarks

When a project lacks a standard and formal ECI system, it will suffer from a shortage of informed construction input into the design phase. However, as discussed above, owners have many options to ensure contractor involvement at all project stages and within the framework of the various PDMs. Key takeaway points on successful ECI include the following:

- Organizations should develop a formal lessons-learned system to capture constructability issues and solutions in a structured format, for use on future projects. Organizations should consider the establishment of such a system as low-hanging fruit that is well worth the effort.
- Owners should hold timely post-construction close-out meetings with contractors to discuss any constructability issues. These meetings are crucial to preventing such problems on future projects.
- While reviewing ATCs proposed by design-builders, owners should not restrict ATC submissions and, when rejecting ATCs, they should begin every response with an explanation of why a proposal was not

accepted rather than simply saying "No". This approach keeps the door open for innovation.

- Owners can benefit greatly from developing a database of contractors who can stand by on an as-needed basis to help with constructability reviews, notwithstanding any organizational or legal constraints.
- Evaluating the environmental impacts of operations and maintenance, not just those of construction, should be an integral part of the constructability process.

Even though this chapter mainly addresses the early inclusion of an external entity (the contractor) in the design process, early design phase involvement of the specialty groups within the organization itself (whether consultants, construction, or maintenance personnel) is critical to project success. If the designers do not solicit input from their own in-house colleagues to improve their designs, the organization loses an important opportunity. Moreover, involving project operations personnel (sometimes from the developer side of Public–Private Partnership projects) in the design phase is a forward-thinking approach to achieving the overall project objectives after turnover.

# About the Authors

**Ghada M. Gad** is an Associate Professor of Construction Engineering in the Civil Engineering Department at California State Polytechnic University, Pomona. She received her Ph.D. in Civil Engineering from Iowa State University and her M.S. and B.S. in Construction Engineering from the American University in Cairo. Her research expertise is in the field of construction project management and has been heavily involved with transportation-related research on the federal and state level, specifically alternative contracting methods, risk management, procurement strategies, and project performance. She has served as PI and Co-PI on NCHRP and ACRP projects, such as NCHRP 25-03: Managing Enhanced Risk in the Mega Project Era, NCHRP 08-112: A Guidebook for Implementing Alternative Technical Concepts into Project Delivery Methods, and NCHRP 24-44: Guidelines for Managing Geotechnical Risks in Design-Build Projects. Having worked for a Bechtel subsidiary through all the construction procurement process stages from contract development, negotiation, administration, and claim analysis, Dr. Gad brings along industry experience specifically tailored to mega construction projects. Dr. Gad has been also actively involved in DBIA and TRB committees. She is currently the communications coordinator of the TRB Committee on Project Delivery and the Research Coordinator of the Contract Law Committee.

**Edward Minchin** is a Professor at the Rinker School of Construction Management at the University of Florida. In a 35-year career, Dr. Minchin has worked as part of eight areas of the construction process — laborer, contractor, designer, inspector, owner, researcher, consultant, and adjudicator. He has 145 publications, including two books and seven book chapters. More of his publications are in the area of project delivery than any other topic. Dr. Minchin currently simultaneously holds three professorships: the UF Term, UF Research Foundation, and Holland Professorships. He was an invited Keynote Speaker at four international Construction Conferences in Panama, China, Canada, and Thailand.

# Chapter 5

# Project Delivery Methods and Green Building Projects: The Owner's Perspective

Bo Xia and Ayokunle Olubunmi Olanipekun

**Abstract**

This chapter is aimed at assessing the preferences of project delivery method selection of green building project owners in the Australian construction industry. Specifically, the project delivery and performance information of 173 Green Star certified projects were obtained for research, where the commonly employed project delivery methods for these green building projects are design–bid–build, design–build and construction management at risk. To assess the preferences of project owners for different types of project delivery methods that foster integration and to test whether the performances of green building projects differed with regard to different project delivery methods chosen by owners, this research shows that the design–build is the most preferred project delivery method by project owners. Additionally, the results revealed that office green building projects where project owners employed the design–build had significantly stronger sustainability performance outcomes. This study contributes to the knowledge on the selection of project delivery methods for green building projects and shows how project owners can contribute to the selection and enhancement of project performance.

**Keywords:** Green building projects, Project delivery methods, Project owners

## Introduction

Green building (GB) has become a notable alternative to the traditional model of physical development in the construction sector. The traditional model has been a long-standing construction process which involves excessive resource consumption and consequently impacts the environment negatively. As of 2018, building construction (and operation) accounted for the largest share of both global final energy use (36%) and energy-related carbon dioxide emissions (39%), rising by 1% from 2017 and 7% from 2010 estimates (IEA, 2019). Also, building materials consumption, especially non-metallic materials such as sand, gravel, limestone and crushed rock, amounted to 35 Gigatonne (Gt) in 2011 and is expected to rise to 82 Gt by 2060 with significant impacts on ozone layer depletion, human toxicity and eutrophication (OECD, 2019). Furthermore, the aggregation of these materials into concrete constituted 9% of global industrial water withdrawal in 2012 and is projected to increase water demand to 75% in regions experiencing water stress (Miller *et al.*, 2018). As an alternative, GB is a resource-efficient option that utilizes less embodied and production energy, materials and water to produce GB projects for healthier and comfortable living (Guan *et al.*, 2020). The GB model of physical development ensures that GB projects are designed, constructed and operated to conserve resources, protect the environment and provide healthy living conditions to people (Ding *et al.*, 2018).

To organize and structure contractual and working relationships among participants involved in the design and execution of green building projects, project delivery methods, such as design–bid–build (DBB), construction management at risk (CMAR), design–build (DB) and integrated project delivery (IPD), should be selected accordingly. It is notable that the relationship is organized and structured differently in different project delivery methods. For instance, the design and construction participants are separated in DBB, while they form a single entity in DB (Kantola *et al.*, 2016).

Different types of project delivery methods promote different levels of integration to maximize GB project performance. Integration could be physical integration, whereby project participants interact directly with one another regarding project issues (Orsi *et al.*, 2020). It could also be timely integration whereby project participants interact frequently to address project issues immediately when they arise. Consequently, integration is a critical element that should be taken into account when selecting any type of project delivery method to achieve high GB project performance (Raouf *et al.*, 2019b). Theoretically, IPD is regarded as promoting the highest level of integration in GB project delivery, followed by DB, CMAR and DBB in descending order of integration levels (see Figure 1). In corroboration, the project delivery methods that promote higher integration are positively correlated with higher GB project performance (e.g. Ibrahim *et al.*, 2020).

Meanwhile, the project owner of green buildings is a stakeholder whose sustainability objectives need to be met and is therefore responsible for determining the type of project delivery methods for GB projects (Olanipekun, Chan, *et al.*, 2017). Since successful GB project delivery is significantly attributed to the appropriate selection of project delivery methods (Raouf *et al.*, 2019a), project owners' preference for the type of

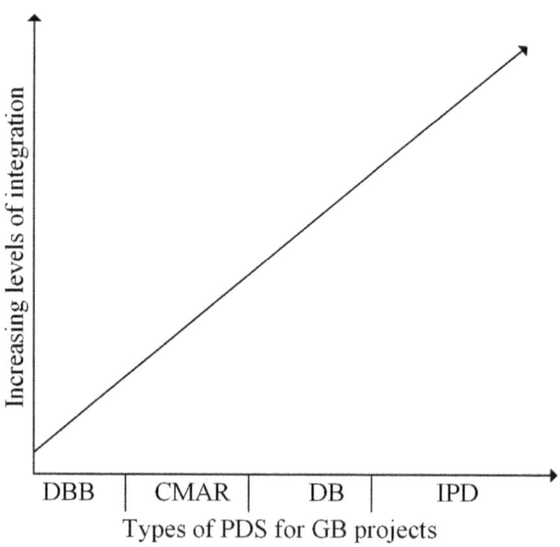


**Figure 1:** Increasing levels of integration in project delivery methods.

project delivery methods that promote integration is therefore important. However, whether the preferences of project owners for project delivery methods promote integration in GB project delivery (Objective 1) is not known, and neither is the impact on project performance (Objective 2). For instance, prior studies have mainly emphasised on the behavior of project owners when they participate in GB project delivery (Ahmad *et al.*, 2019; Zhang *et al.*, 2019). However, the behaviors indicate neither the preferences of project owners for different project delivery methods nor the impact of the preferences on GB project performance. Also, another study that evaluated GB project performance of different types of project delivery methods using data submitted by contractors (Ibrahim *et al.*, 2020) neglected the perspective of project owners and its impact on GB project performance remains unknown. The cross-pollination of both objectives will fill this gap.

This chapter is aimed at assessing the project delivery method preferences of GB project owners in the Australian construction industry and providing answers to the two objectives. Since 2004, the Green Building Council of Australia (GBCA) has certified two thousand, nine hundred and twenty six (2926) GB projects for different types of project owners (Green Building Council of Australia, 2020). The office, education, retail, residential, public, industrial and health GB project owners often employ CMAR, DBB and DB for GB project delivery in the country. The IPD is still an unorthodox project delivery method (Kantola *et al.*, 2016) and has not been implemented widely for GB project delivery in Australia (Raouf *et al.*, 2019a). The IPD was therefore not considered in the current study. This chapter makes a significant contribution by providing a pathway for project owners to enhance GB project delivery through their preferences for project delivery methods in the Australian construction industry. As the types of GB project ownership are universally similar, the pathway is useful for GB project delivery in other countries as well.

## *Project owners' involvement in green building project delivery*

Project owners' active participation contributes to the successful delivery of green building projects. Existing research has mainly focused on behavioral aspects of project owners' participation. For instance,

Olanipekun, Chan, *et al.* (2017) and Zhang *et al.* (2019) focused on project owners' participation through owner commitment in the delivery of green building projects. Basically, owner commitment involves the active and deliberate participation of project owners in the delivery of green building projects, and the indicators of owner commitment include but are not limited to facilitating the integration of project participants and educating them to achieve sustainable solutions (Olanipekun, Chan, *et al.*, 2017). Project owners' motivation, which is classified into intrinsic and extrinsic ones (Olanipekun, Xia, *et al.*, 2017), is necessary to increase their commitment toward the successful delivery of green building projects (Zhang *et al.*, 2019). Project owners could be intrinsically motivated out of personal or individual volition (such as pro-environmental altruism) or extrinsically motivated to benefit separable rewards (or avoid punishments) (Iheanyichukwu Joachim *et al.*, 2015). Interestingly, Olanipekun, Xia, *et al.* (2017) found that project owners' intrinsic motivation was more important than the extrinsic one, and also, both complemented one another in driving project owners' delivery of green building projects in the Australian construction industry. Furthermore, Olanipekun *et al.* (2018) found that the intrinsic motivation of project owners was required to increase their commitment, which consequently enhanced the sustainability performance of green building projects.

Meanwhile, a recent study by Raouf *et al.* (2019a) concluded that the successful delivery of green building projects in the construction industry is attributed to the choice of project delivery methods, which structure the relations and responsibilities among stakeholders involved in project delivery (West *et al.*, 2012). The project delivery method not only structures the relations and responsibilities among stakeholders and processes involved in green building but also integrates them to maximise the performance of green building projects (Ahmad *et al.*, 2019; Orsi *et al.*, 2020). Therefore, employing the appropriate project delivery method that integrates the interdisciplinary network of stakeholders and project processes enhances the performance of green building projects (Gunhan *et al.*, 2012; Raouf *et al.*, 2019a). As summarised in Table 1, integration within types of project delivery methods differs, and project delivery methods that promote higher integration result in higher GB project performance compared to those with less integration. From Table 1, it could

**Table 1:** Summary of literature on green building (GB) project delivery methods (PDMs).

| | Year | DBB | DB | CMAR | IPD | Performance criteria | Unit of analysis | Location | Method | Remarks |
|---|---|---|---|---|---|---|---|---|---|---|
| P1 | 2020 | x | x | | | Analysis of time, cost and level of sustainability | Health and Office building projects | Italy & Spain | Quantitative analysis of case projects | DB fosters process integration which enhances the performance of GB projects |
| P2 | 2020 | x | x | x | x | Cost, schedule, quality, safety, communication, and change management | Industrial, commercial, infrastructure and multi-storey residential project information from contractors | USA, Canada, Colombia, Ireland | Quantitative analysis of project information | IPD and collaborative PDS such as DB produce better GB performance than DBB. This remark is limited to traditional projects. |
| P3 | 2019 | | x | | | Analysis of cost of quality | Residential building project | Qatar | Quantitative analysis of a case project | Constructability feedback is more viable in DB methods. It increases stakeholders influence on cost of quality |
| P4 | 2019 | Traditional performance criteria (cost, quality and schedule) and sustainability performance (social, economic and environmental) | | | | Perceptions of GB experts | Australia, Singapore, India, UAE, Hong Kong, UK & Pakistan | Questionnaire survey & interview | PDS affect the association among GB performance criteria | |

(*Continued*)

**Table 1:** (*Continued*)

| | Year | DBB | DB | CMAR | IPD | Performance criteria | Unit of analysis | Location | Method | Remarks |
|---|---|---|---|---|---|---|---|---|---|---|
| P5 | 2019 | x | x | x | x | Traditional performance, social, economic and environmental sustainability, and participant satisfaction | Conceptual variables | NA | System dynamics modelling of variables | IPD and collaborative PDS such as DB produce better GB performance than DBB |
| P6 | 2014 | x | x | x | x | Traditional and sustainability performance | Net zero energy education building project | Finland | Expert interview of case project | IPD accompanied with competitive negotiations is a more suitable delivery method for zero energy building project. The CMAR accompanied with IPD target pricing and competitive negotiation can be very suitable. |
| P7 | 2011 | x | x | x | | Analysis of time, cost, quality and level of high performance (e.g. energy and IEQ) | Office building projects | USA | Quantitative analysis of case projects | CMAR and DB result in higher overall success than DBB |
| P8 | 2010 | x | x | x | | Analysis of cost growth, construction speed, delivery speed and energy rate | Office building projects | USA | Quantitative analysis of case projects | In terms of delivery speed, DB and CMAR outperform DBB |
| P9 | 2010 | | x | | | Analysis of sustainability rating | Office, Health & Education building projects | USA | Content analysis of building procurement documents | DB green building projects procured using best-value procurement achieves higher sustainability ratings |

*Notes*: P1: Orsi *et al.* (2020); P2: Ibrahim *et al.* (2020); P3: Raouf *et al.* (2019b); P4: Ahmad *et al.* (2019); P5: Tang *et al.* (2019); P6: Kantola *et al.* (2016); P7: Korkmaz *et al.* (2011); P8: Korkmaz *et al.* (2010); P9: Molenaar *et al.* (2010).

be seen that DB and CMAR promote higher stakeholder and process integration (Kantola *et al.*, 2016; Tang *et al.*, 2019), while the integration within the DBB is very minimal (Molenaar *et al.*, 2018). Comparatively, integration within the IPD is the highest and resulted in the highest GB project performance (Ibrahim *et al.*, 2020). To sustain high GB project performance, the study recommended that the construction industry should move toward IPD and away from DBB.

Clearly, integration is a critical element that should be considered in determining the best project delivery method for successful GB project delivery. At the same time, project owners are the primary stakeholders who determine the project delivery method for GB projects (Kantola *et al.*, 2016). It means that project owners need to be mindful of the critical element of integration when choosing project delivery methods for GB projects to ensure successful project delivery. Therefore, the preferences of GB project owners toward project delivery methods are important. For instance, there are different types of GB project owners according to project uses, such as the office, residential, industrial, retail and health projects in the Australian construction industry (Olanipekun *et al.*, 2018). In this case, the GB project delivery method preferences should be explicit for any type of project owner according to different types of GB projects. Furthermore, GB project performance, which comprises traditional (cost, time and quality) and sustainability performances (e.g. sustainability rating and energy measures), vary when different project delivery methods are employed (Ibrahim *et al.*, 2020; Raouf *et al.*, 2019a). Therefore, the type of project delivery methods selected by project owners is essential to successful GB project delivery.

Raouf *et al.* (2019a) identified that the commonly employed project delivery methods for green building projects in the construction industry include design–bid–build (DBB), design–build (DB) and construction management (CMAR). DBB is a traditional method which unmerges the design and construction phases and the project owner commissions the designer to complete the design which is consequently used to solicit for contractor's bid and actual project execution (Tang *et al.*, 2019). DB allows the design and construction phases to merge because the owner has a contractual relationship with one entity — a design–build contractor who provides design and construction services with the possibility of subcontracting parts of the work (Molenaar *et al.*, 2018). CMAR is a system where the owner commissions a construction manager at risk to deliver a project within a Guaranteed Maximum Price (GMP) which is based on the project documents and specifications at the time of the GMP plus any reasonably inferred items or tasks (Ibrahim *et al.*, 2020). Normally, the CMAR gives the owner a GMP before bidding, and by doing so, assumes the risk of bids coming in higher as the CMAR is contractually bound to deliver the project within the defined GMP.

## Hypothesis development

GB projects are complex due to the technologies that are integrated, and the multidisciplinary expectations required to conform to their specifications (Raouf *et al.*, 2019a). As a result, GB projects should be delivered using the project delivery method that promotes integration to overcome the complexities and increases GB project performance. At the same time, the project owner is the stakeholder primarily responsible for determining the type of project delivery methods for GB project delivery. It means that the preference of project owners for project delivery methods that promotes integration may impact (or not) GB project performance. Therefore, this chapter hypothesizes as follows.

$H_A$: *GB project performance differs between different types of project delivery methods preferred by GB project owners.*

$H_0$: *GB project performance does not differ between different types of project delivery methods preferred by GB project owners.*

As illustrated in Table 1, there are many criteria for evaluating the performance of GB projects. In most cases, these criteria are used in quantitative analysis of the cost, time and level of GB project sustainability (e.g. P1). This involves the collection of real project data and the determination of the differences between the expected and obtained results. In a few cases, the criteria are rather subjective such as obtaining expert opinions of the performance of completed GB through an interview (P4) and dynamic modelling (P5). To evaluate the performance of GB projects in this study, the sustainability and traditional performance criteria are considered (Ahmad *et al.*, 2019; Raouf *et al.*, 2019b). The sustainability performance evaluates the level of greenness in GB projects, while the traditional performance evaluates the efficiency of project delivery in terms of cost, time and quality. Both criteria produce a comprehensive assessment of GB project performance in the construction industry (P6).

## Research methodology

This chapter employs a quantitative research methodology comprising a questionnaire survey and data analysis to investigate the project delivery methods preferences of green building project owners. Therefore, GB project owners in the Australian construction industry were the unit of analysis. In Australia, the practice of GB is professionally regulated by the Green Building Council of Australia (GBCA) to ensure maximum GB project delivery performance. The members of the GBCA known as the Green Star Accredited Professionals (GSAPs) are accredited to represent the interests of project owners in the delivery of GB projects in the country. In other words, the GSAPs are employed by project owners to execute the delivery of GB projects. The GSAPs were used as a proxy for GB project owners to respond to the questionnaire.

The questionnaire had three sections. In the first section, the GSAPs were asked questions about the professional capacities in which they work for GB project owners such as primary roles and number of years of experience. In the second section, the GSAPs were asked to recollect the last GB project executed on behalf of project owners and provide information about the types of projects, types of ownership and the types of project delivery methods employed. In the third section, the GSAP were asked to score the level of performance achieved in the recollected GB projects. Similar to the recent study of Ahmad *et al.* (2019), the GSAPs were provided Likert scale options from *very low* (1) to *very high* (5) to score the performance. The performance criteria that formed the basis of scoring include the traditional and sustainability ones. The former was subdivided into *completeness of project within targeted budget, amount of callbacks or reworks for defective work, completeness of project within the scheduled time*. Similarly, the latter was subdivided into *level of sustainability rating achieved on the project, level of energy efficiency in the building project and quality of indoor air of the project.*

Using an institutional web tool, the survey questionnaire was developed into an online format and sent to the emails of all the GSAPs listed on the GBCA website (https://www.gbca.org.au/people/?) while reminders for completion were sent on two occasions within a period of six months after which the survey was closed. With the aid of the web tool, a qualifying question that enabled only the GSAP who have been involved in GB project delivery on behalf of project owners to respond to questions in the three sections was introduced. One hundred and seventy-three (173) GSAPs responded, and the responses regarding their individual capacities and GB project information were analyzed. The methods of data analysis comprised percentage frequency analysis and the Kruskal–Wallis test. The percentage frequency distribution was used to specify the percentage of responses given by GSAPs about their professional capacities and the projects they executed on behalf of project owners (Shapiro, 2008). The premise of the decision was based on the responses with the highest frequency or percentage(s). Also, the Kruskal–Wallis test was used to evaluate if there were differences in the GB performance score (dependent variable) between the types (or groups) of project delivery methods (e.g. CMAR, DBB, DB) (independent variable). It is a non-parametric test for determining if groups' medians are statistically significantly different (Chan *et al.*, 1997). Also, the distributional assumption of the test was evaluated by the visual inspection of the similarity in the shape of the distributions in each of the types of project delivery methods in the box-plot chart (Frigge *et al.*, 1989). The results were interpreted based on the statistical significance of the median (M) of GB project performance scores between the types of project delivery methods, the *H*-statistic (or $\chi^2$-distribution) and degrees of freedom (df). $H_0$ with $p$-value > 0.05 was retained, and the vice versa rejected on account of statistically significant group. To know which groups differ from each other, the Dunn (1964) procedure with Bonferroni adjustment was used *post-hoc*.

## Results

### *Data description*

The results of the primary roles of the GSAP revealed that 49% of them are environmentally sustainable design (ESD) consultants, followed by construction planning and execution team members (32%) such as builders and project managers, and the least are designers and architects (18%). The GSAPs (42%) have been consulting for GB project owners: for up to five years (31%), for more than five years but not 10 years (42%), and for over 10 years (27%). Majority of them (69%) are well-experienced consultants for GB project owners. The results revealed 173 number of GB projects executed by the GSAP on behalf of project owners in total. 47% ($n$ = 81) were office GB projects, followed by education GB projects (19%, $n$ = 33), residential GB projects (16%, 27) and public GB projects (8%, $n$ = 13). The rest in descending order were retail GB projects (5%, $n$ = 8) and health and industrial GB projects (3%, $n$ = 5, respectively). Finally, all the GB projects were Green Star rated: Six Star rating or World leadership (52%, $n$ = 83), Five Star rating or Australian excellence (42%, $n$ = 67) and Four Star rating or best practice (7%, $n$ = 11).

## *Main results*

In terms of the primary use of the 173 GB projects, they were owned by office, education, residential, public, retail, health and industrial GB project owners in the Australian construction industry. The result of the project delivery method preferences of project owners is presented in Table 2. The DB was used for the delivery of 87 GB projects, followed by DBB (47 GB projects) and lastly, the CMAR (39 GB projects). The result suggested that DB is the most preferred project delivery method among GB project owners. When splitting the number of DB projects by types of projects, 43 are office buildings, 17 residential buildings, 5 retail buildings and 5 industrial buildings. The DBB was used for the delivery of the highest number of education ($n$ = 12) and public ($n$ = 5) GB projects, while the DB and CMAR were equivalently used for the delivery of the highest number of health ($n$ = 2) GB projects.

The Kruskal–Wallis $H$-test was used to test the hypothesis that GB project performance scores differ between CMAR, DBB and DB. GB project performance was divided into the traditional and sustainability performances, and their scores were obtained for different types of GB projects. The visual inspection of the boxplot showed that the GB project performance scores (traditional and sustainability) for the three types of

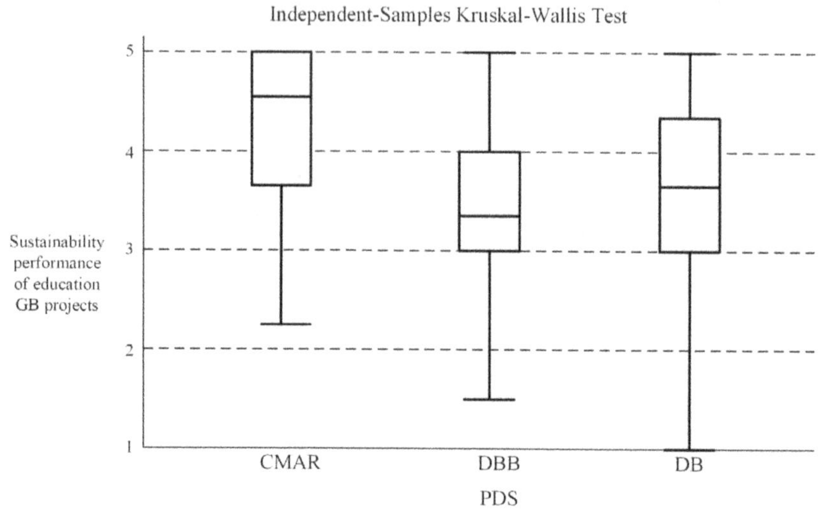


**Figure 2:** Boxplot chart of median scores distribution of education green building projects.

project delivery methods were similar in shape (or similarly distributed) which permitted their medians (*M*) to be compared. Due to space consideration, the boxplot for the sustainability performance scores of education GB projects is presented in Figure 2, and it shows close similarity in the distribution of performance scores. However, the industrial, retail and health GB project performance scores for project delivery methods were dissimilar in shape and therefore discontinued from further analysis. The industrial ($n = 6$), retail ($n = 5$) and health ($n = 5$) GB projects had very small sample sizes for the groups (e.g. CMAR, DBB, DB) which affected the shape of their distribution (Chan *et al.*, 1997).

Tables 2 and 3 report the statistical results of the median of GB project performance scores between the types of project delivery methods. The median traditional performance scores for education ($\chi^2(2) = 1.541$, $p = 0.463$), office ($\chi^2(2) = 0.833$, $p = 0.659$), public ($\chi^2(2) = 1.293$, $p = 0.524$) and residential ($\chi^2(2) = 3.358$, $p = 0.187$) GB projects were not statistically significant between CMAR, DBB and DB. Similarly, the median sustainability performance scores for education ($\chi^2(2) = 3.329$, $p = 0.189$), public ($\chi^2(2) = 2.079$, $p = 0.354$) and residential ($\chi^2(2) = 0.040$,

**Table 2:** Kruskal–Wallis test results for traditional performance.

| Types of GB projects | Traditional performance of education GB projects | | | | Traditional performance of office GB projects | | | | Traditional performance of public GB projects | | | | Traditional performance of residential GB projects | | | |
|---|---|---|---|---|---|---|---|---|---|---|---|---|---|---|---|---|
| **Project delivery methods** | **CMAR** | **DBB** | **DB** | **Total** | **CMAR** | **DBB** | **DB** | **Total** | **CMAR** | **DBB** | **DB** | **Total** | **CMAR** | **DBB** | **DB** | **Total** |
| $N$ | 10 | 12 | 11 | 33 | 17 | 21 | 43 | 81 | 4 | 5 | 4 | 13 | 3 | 7 | 17 | 27 |
| $M$ | 4.00 | 3.67 | 4.00 | 4.00 | 4.00 | 3.67 | 4.00 | 4.00 | 4.00 | 3.67 | 3.67 | 4.00 | 5.00 | 3.00 | 4.00 | 3.67 |
| $p$-value | 0.463 | | | | 0.659 | | | | 0.524 | | | | 0.189 | | | |
| $H$ | 1.541 | | | | 0.833 | | | | 1.293 | | | | 3.358 | | | |
| df | 2 | | | | 2 | | | | 2 | | | | 2 | | | |

**Table 3:** Kruskal–Wallis test results for sustainability performance.

| Types of GB projects | Sustainability performance of education GB projects | | | | Sustainability performance of office GB projects | | | | Sustainability performance of public GB projects | | | | Sustainability performance of residential GB projects | | | |
|---|---|---|---|---|---|---|---|---|---|---|---|---|---|---|---|---|
| **Project delivery methods** | **CMAR** | **DBB** | **DB** | **Total** | **CMAR** | **DBB** | **DB** | **Total** | **CMAR** | **DBB** | **DB** | **Total** | **CMAR** | **DBB** | **DB** | **Total** |
| $N$ | 10 | 12 | 11 | 33 | 17 | 21 | 43 | 81 | 4 | 5 | 4 | 13 | 3 | 7 | 17 | 27 |
| $M$ | 4.84 | 4.17 | 4.33 | 4.33 | 3.25 | 4 | 4.00 | 4.00 | 5.00 | 4.33 | 4.17 | 4.33 | 3.67 | 4.33 | 4.00 | 4.00 |
| $p$-value | 0.189 | | | | 0.000** | | | | 0.354 | | | | 0.989 | | | |
| $H$ | 3.329 | | | | 19.98 | | | | 2.079 | | | | 0.04 | | | |
| df | 2 | | | | 2 | | | | 2 | | | | 2 | | | |

**Table 4:** *Post-hoc* test results and number of retail, health and industrial GB projects.

| Sample 1–sample 2 | Test statistic | Std. error | Std. test statistic | Sig. | Adj. sig. |
|---|---|---|---|---|---|
| CMAR–DBB | –27.174 | 7.603 | –3.574 | 0.000 | 0.001 |
| CMAR–DBB | –28.947 | 6.677 | –4.335 | 0.000 | 0.000 |
| DBB–DB | –1.773 | 6.204 | –0.286 | 0.775 | 1.000 |

$p = 0.989$) GB projects were not statistically significant between CMAR, DBB and DB. Therefore, $H_0$ with $p$-value $> 0.05$ was retained. In contrast the median sustainability performance scores for office GB projects was statistically significant between CMAR, DBB and DB ($\chi^2(2) = 19.980$, $p = 0.000$). Therefore, $H_0$ with $p$-value $< 0.05$ was rejected. Subsequently, the Dunn (1964) procedure with Bonferroni adjustment *post-hoc* test was used to identify which of CMAR, DBB and DB were statistically different from one another. As presented in Table 4, the result revealed statistically significant differences in median sustainability performance scores between CMAR ($M = 3.25$) and DBB ($M = 4.00$) ($p = 0.001$), and CMAR and DB ($M = 4.00$) ($p = 0.000$), but not between DBB and DB ($M = 4.00$).

## *Discussion of findings*

This chapter examines the frequency of use across different project delivery methods in 173 GB projects based on the responses of project owners. The analysis revealed that the DB, DBB and CMAR were used in eighty seven (87), forty seven (47) and thirty nine (39) GB projects, respectively. It could be seen that the DB was the most frequently used and therefore the most preferred project delivery method by GB project owners. It is notable that, while some recent studies have explored stakeholder perspectives on project delivery methods for GB projects, none revealed the preferences of GB project owners. Ferme *et al.* (2018) examined the perspectives of 37 Green Star professionals in the Australian construction sector and revealed their preferences for the DB and Management Contracting (MC) delivery methods because both promote collaboration among them. Also, Gunhan (2019) examined the perspectives of major green builders in the USA and revealed that the DB was their preferred project delivery method for GB projects. As a primary stakeholder in GB project delivery, the revelation of the preferred type of project delivery methods of GB project owners contributes to the existing literature. As the primary stakeholder in GB project delivery (Kantola *et al.*, 2016), project owners shall select the appropriate type of project delivery methods for GB projects.

Additionally, this chapter tests the hypothesis of whether GB project performance differed between different types of project delivery methods preferred by GB project owners. This would in turn help GB project owners to select the specific types of delivery methods for maximizing GB project performance (against both traditional and sustainability performance criteria). Past studies have evaluated GB project performance without linking the performance outcomes (e.g. cost and schedule outcomes) to the types of project delivery methods employed (Hwang *et al.*, 2015; Hwang *et al.*, 2017). This gap has been filled in recent studies whereby the evaluation of GB project performance was linked to the types of project delivery methods. The hypothesis testing in this chapter reveals that the difference of the traditional and sustainability performances for education, public and residential GB projects was not statistically significant between CMAR, DBB and DB. Therefore, $H_0$ was retained. It meant that the traditional and sustainability performances of education, public and residential GB projects did not differ against the project delivery preferences of GB project owners. In contrast, the median sustainability performance for office GB projects was statistically significantly different between CMAR, DBB and DB. More specifically, the sustainability performance of office GB projects differed with regard to the project delivery preferences of GB project owners. In comparison with education, public and residential GB projects, office GB projects need to meet high sustainability requirements to be commercially viable (Eichholtz *et al.*, 2013). Therefore, as demonstrated in Orsi *et al.* (2020), office GB project owners need to select the project delivery method that fosters integration to ensure high sustainable GB project performance. Therefore, in this study, the office GB project owners have a high preference for the DB method ($n = 43$) as it fosters high integration.

Furthermore, there exists a statistical difference in the sustainability performance of office GB projects based on the project delivery preferences of GB project owners. A *post-hoc* test was performed which revealed that both the DB and DBB methods produced the equivalent and higher office GB project sustainability performance scores ($M$ = 4.00) than CMAR ($M$ = 3.25). Although it is notable that the DB method provides a higher level of integration for maximizing GB project performance (Molenaar *et al.*, 2010) compared to the DBB method (Orsi *et al.*, 2020), the findings in this chapter show that project owners can select either the DB or DBB method to maximise the sustainability performance of office GB projects. Meanwhile, it is interesting to note that the DBB method produced high sustainability performance scores, which is in contradiction to many previous studies. The DBB method is where the design and construction phases are unmerged (Tang *et al.*, 2019), thereby preventing process integration of sustainability features and tasks by project participants (ORSI *et al.*, 2017). The underlying reasons remain to be explored further.

## Conclusion

In summary, this research assesses the project delivery preferences of project owners in the Australian construction industry. The commonly employed project delivery methods for green building projects are design–bid–build, design–build and construction management at risk, and they foster different levels of integration in practice. This research shows that the design–build is the most preferred by GB project owners. Furthermore, the hypothesis on whether the performances of different types of green building projects differed based on the project delivery method preferences of project owners was tested. The results revealed that office green building projects where project owners employed the design–build method had significantly stronger sustainability performance outcomes.

This study contributes to the knowledge on the selection of appropriate project delivery methods for green building projects. The current literature is flooded with knowledge on how other stakeholders (such as designers, contractors and consultants) contribute to selecting project delivery methods for green building projects. This study shows how project owners can contribute to the selection of project delivery methods for green building projects and also indicates that there is a link between the project delivery method preferences of project owners and the performance of green building projects. The project owners' preferences for design–build can lead to a high sustainability performance of green building projects.

# About the Authors

**Bo Xia** is an Associate Professor in the School of Architecture and Built Environment, Queensland University of Technology. His research interests are in design-build, construction procurement, sustainable buildings, and built environment for older people. His research findings have been published in internationally-recognized journals. Currently, Aspro. Xia acts as the regional editor of International Journal of Construction Management, member of Editorial Boards for the ASCE Journal of Management in Engineering.

**Ayokunle Olubunmi Olanipekun** is a Senior Lecturer in Quantity Surveying in the School of Architecture and Built Environment, University of Wolverhampton, Wolverhampton, England, UK. He has professional academic experience in higher education institutions in Nigeria, Australia and New Zealand since 2011. He is Fellow of the UK Higher Education Academy and a Professional Member of the Nigerian Institute of Quantity Surveyors (NIQS).

# Part B

# Procurement

# Chapter 6

# Procurement Approaches for Design, Construction, and Materials

Pramen P. Shrestha[1] and Giovanni C. Migliaccio[2]

[1]*University of Nevada Las Vegas*
[2]*University of Washington*

## Procurement overview

In Chapter 1, we introduced the concept of project delivery method as the overall framework for delivering a project. However, this concept alone does not fully describe project delivery. Indeed, a project's execution is substantially affected by two additional steps: procurement and contracting. Built environment projects are delivered by a myriad of project parties, including numerous specialty designers as well as general and specialty contractors. All these parties procure the project and perform their duties according to their either prime contracts or subcontracts. Figure 1 provides the graphical depiction of a hypothetical project, where the project owner has contracted design and construction services separately and several subcontracts are in place. A prime contract regulates the direct relationship between the project owner and another project entity. Subcontracts, instead, are all the other contracts that do not involve the project owner. Subcontracts often incorporate language similar to their overarching prime contract as a way to promote alignment to the overall project objectives. However, each one of these contracts may have been procured with a different procurement approach.

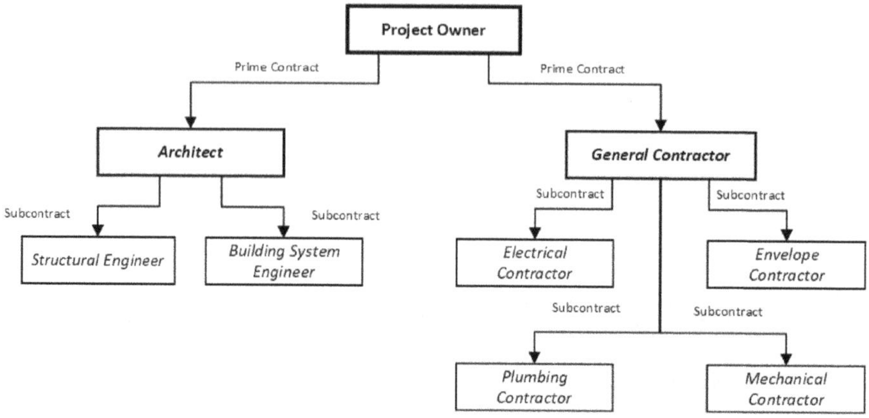


**Figure 1:** Design-Bid-Build Project with Agency Construction Manager.

This book relies on a specific definition of procurement as the act of purchasing external services and materials necessary to deliver a project. When purchasing services whose scope will be defined in a prime or subcontract, the purchasing party will procure a service provider. In the example in Figure 1, the project owner has separately procured an architect and a general contractor for the project, who have later procured a series of specialty designers and subcontractors to help them achieve the project objectives.

## Procurement selection methods

Selecting the procurement approach means making a series of decisions that will affect the process and its outcome. To start with, identifying what decision criteria will guide the selection will clearly impact procurement. Project owners can broadly identify decision criteria that align with the project objectives; still, it is common that most criteria will revolve around the usual dimensions of success for built environment projects, namely cost, time, quality, and safety, weighted differently from project to project. Secondly, the selection process can encompass a single step or multiple steps. Moreover, some project owners may consider some project components to be particularly important; therefore, they may request competing general contractors (GC) to propose the subcontractors for these components and then evaluate each GC together with their team. In the following subsections, several procurement methods are described, together with their underlying selection decision criteria and selection processes.

## *Low bidding*

Using price as the main criterion to select construction firms, low bidding is frequently used to procure services under design–bid–build delivery, by identifying prime contractors and often subcontractors. While design–bid–build and low bidding are still prevalent in the public realm to select the prime contractor and often subcontractors, an increased use of different delivery methods on public projects has resulted in the adoption of different procurement approaches, which are discussed later in this chapter.

While price is sometimes used to select providers of design service, the use of price-based competitive bidding is rarely used to award design contracts in the United States. One of the main reasons for not using low bidding on design procurement derives from the nature of design–bid–build, where design documents are developed early on by designers before being incorporated into the procurement documents used as the basis for soliciting bids from competing construction firms. Competing firms use these documents as the basis for estimating project cost and evaluating their risks, so they can add their markup based on their risk considerations and submit a bid price to the project owner. The submission of a bid often legally binds a construction firm to signing the contract at the bid price if its bid is selected. The selected prime contractor is, however, still entitled to request a contract change by negotiating a change order with the project owner under various design-dependent scenarios, such as the design is flawed or needs to be changed later. This entitlement acts as a disincentive for owners to purchase design services based on the price alone or mainly on price.

Competitive low bidding is popular because it is simple for a public agency to implement and for the public to understand. However, using only the price to select a construction firm has several disadvantages; several variations to standard low bidding exist to address these disadvantages.

As producing a set of plans and specifications that can be built exactly as designed to the detailed level is basically impossible, asking contractors to price a project in which many uncertainties still exist in the procurement documents is known to trigger various issues, including conflicts during the contract administration and price escalations through contract changes. Examples of issues include differing site conditions that force the redesign of entire elements of the scope or constructability issues that can only be detected through design construction review and coordination. To mitigate these issues, owners may contractually mandate specific processes that force the contractor to identify these issues in a timely manner and discuss them with the project owner, while also forcing them to record associated costs to owner's contingencies funds, so that the contractor's bid and change order requests do not hide their actual costs.

Additionally, the selection of a construction firm on price alone does not allow owners to address other tangible and intangible factors that affect project success, with the unintended consequence of selecting a firm that is not able to achieve some of the project objectives, other than the budget, or even to complete the project at all. As the rise of these situations may be negative, if not devastating, for a project, owners often modify purely low bidding. To allow only reliable contractors to submit a bid, some owners may pre-qualify firms before inviting them to submit a bid. While the bid winner would still be the low bidder, only pre-qualified firms would be allowed or invited to submit a bid. Some owners maintain a list of pre-qualified firms for each scope item, from general to specialty construction and update it periodically. An alternative to directly pre-qualify contractors is to incorporate bonding requirements in the procurement by requesting competing contractors to submit their bid together with a bond. This approach outsources the investigation on the reliability of the competing construction firms to complete the given project to sureties. The sureties would then take specific risks such as:

- The contractor's inability to follow-up on a bid through the issuance of a bid bond.
- The failure of the firm to perform through issuance of a performance bond.

- The failure to pay subcontractors and vendors through the issuance of a payment bond.

For projects where early completion is beneficial and preferred, owners may adopt a price and time selection, where competitors need to submit a project duration along with their bid price. Under this approach, each day of project duration is assigned a monetary value, and the bid price would be the sum of the estimated price plus the time monetary value. Last, price selection is often seen as a barrier to construction innovation that can lead to a better built project. A common approach to address this is to incorporate contractual incentives for construction firms to propose changes to the design package that would add value or reduce lifecycle costs, also known as value engineering proposals during contract execution.

## *Best-value selection*

Competing firms are asked to develop project proposals, which are evaluated against their ability to meet the project owner's objectives through a multi-criteria evaluation, which includes quantitative criteria, such as price or fee percentage, target project duration, and safety performance indices, and qualitative criteria, such as references, experience with the project type, and experience and qualifications of the proposed project staff, including the project manager and superintendent. Best-value selection is more flexible, as it can be adapted for use under any project delivery method.

Best-value selection usually follows either a one-step or two-step selection process. Under the one-step best-value process, competing firms are invited to respond to a request for proposal (RFP) that details the expected content of the proposals and describes the selection process. The one-step approach is used on smaller or simpler projects when the information to be submitted is limited or competition is expected to be scarce. In a one-step procurement, proposers submit both technical and price proposals to the owner in a single package.

In all other situations, a two-step approach is preferred. Under the two-step best-value selection, competing firms are first invited to respond to a request for qualification (RFQ) by submitting information on their firms that would help the project owner pre-qualify firms based on their chances to achieve all of the project objectives at the best value. Then, shortlisted firms are invited to respond to an RFP, where each firm will develop their full proposal for delivering the project. Sometimes, the owner will ask contractors to submit the price proposals in a sealed bid that will be opened once the technical proposals have been evaluated — to avoid bias in the process. Some owners prefer to sequence the submission and initially ask for an unpriced technical proposal from contractors; then, they invite only those contractors who the owner deems as technically qualified to submit sealed bids. In these cases, the contract is awarded to the lowest bidder that the owner technically qualifies.

An alternative to the two-step approach described above is a process where the owner, in the first step, evaluates the proposers' qualifications submitted in response to an RFQ. Then, the owner will pre-qualify, instead of shortlist, proposers. Under a pre-qualification process, the owner will move all the proposers meeting the minimum criteria to the next phase, whereas in a shortlist process, the owner invites a small number of the most qualified proposers, generally three to five, to move on to the next phase. The second step then involves the remaining proposers responding to a separate RFP by submitting both a technical proposal and a separate price proposal, with the result most often being the selection of the proposer that provides the best value to the owner. The approach to evaluate technical and price components is crucial to the result, as discussed in a later chapter.

The two-step procurement process allows an owner to select a contractor based on cost and other factors. This best-value selection process ensures that owners award contracts to qualified contractors that have submitted an acceptable technical proposal and a reasonable bid price. During the contractor procurement phase, an owner can use an independent third party estimating consultant to check the project costs and bids by the proposers. This type of selection process can be used to select contractors, independent of the alternative delivery method used, including the selection of construction managers (CMs) under CMAR, design-builders under lump sum DB or progressive DB, and developers under public–private partnership (PPP).

While best-value procurement could be used on any type of project delivery, it is mainly popular in delivery methods other than design–bid–build.

### *Qualifications-based selection*

Qualifications-based selection is highly flexible and can be adapted for use under any project delivery method. However, this selection process deemphasizes price by adopting a multi-criteria selection, where the price criterion is minimally weighted, if considered at all. This characteristic reduces its use in selecting contractors when design is in advanced stages of development.

### *Sole source selection*

Among all the discussed procurement approaches, sole source selection is the only approach that excludes a somehow open competition process. Under sole sourcing, a service or material provider is instead selected because they are deemed to be the only one capable of providing the service or are preferred to satisfy the project needs for justifiable reasons, so a competition is not needed. This is a common approach for selecting specialized project components, such as building equipment manufacturers and niche specialty contractors. In the public sector, the decision to use this non-competitive approach often requires justification, including investigations of potential competitors or prior good references. Different from previously described procurement approaches, the burden of this data gathering would be on the selecting party. While its use in the public sector is strongly limited and constrained, sole sourcing is much more freely adopted in the private sector.

## Summary

As most owners will not self-perform projects, procuring design and construction services is an important step to guarantee the delivery of a project, and procurement affects success. This chapter was just a brief introduction to the topic of procurement of design and construction services. The forthcoming chapters will continue to build upon these concepts by elaborating more on the issues that are critical in understanding the project delivery methods. This section of the book covers four chapters related to project delivery. Chapter 7 provides more details on qualifications-based procurement. Chapter 8 reports on the fallacy of best-value procurement, where high quality technical proposals are often not rewarded, as the selection algorithms are often biased toward price. The final Chapter, 9, reports on the technical and organizational challenges for incentivizing sustainable public procurement.

# About the Authors

**Pramen P. Shrestha** is a Professor of Construction Engineering Management and Graduate Coordinator in the Civil & Environmental Engineering and Construction department at the University of Nevada, Las Vegas (UNLV). He holds an M.Sc. in Civil Engineering from Oklahoma State University and a Ph.D. in Civil Engineering from the University of Texas at Austin, United States (US). He completed his B.Sc. in Civil Engineering from the National Institute of Technology, Jamshedpur, India. He had about 10 years of professional experience in design and construction of highway and infrastructure projects in Nepal before moving to US. After completing his Ph.D. in Civil Engineering from the University of Texas at Austin in 2007, he joined UNLV as an Assistant Professor. He has received federal, state, and private research grants worth about $2.5 million. His research grants are funded by the US State Departments of Transportation, Transportation Research Board, National Science Foundation, US Labor Department, Water Design Build Council (WDBC), Airport Cooperative Research Program (ACRP) and private agencies. He has helped with the graduation of five Ph.D. and 34 master's students as an advisor. He currently has two Ph.D. students and three master's students working for him on various projects. He has published more than 55 peer-reviewed journal papers and 75 conference proceedings. He is a Fellow of American Society of Civil Engineers and a licensed professional engineer in Texas.

**Giovanni C. Migliaccio** holds a Howard S. Wright Endowed Professorship, he is the current Chair of the Department of Construction Management, and he has served as the Executive Director of the Center for Education and Research in Construction (CERC). He holds an M.S. and a Ph.D. in Civil Engineering from the University of Texas at Austin and a master-level degree from Politecnico di Bari in Italy. Prior to moving to the US, he worked in Italy in the construction management of telecommunication projects under Nortel Networks, Nokia Networks, and IPSE 2000. At UW, Giovanni is active in all three pillars of academic life, including teaching, research, and service to the university and the industry.

# Chapter 7

# Qualifications-driven Procurement

Mounir El Asmar and Wylie Bearup

*Arizona State University*

## Abstract

Qualifications-driven procurement has been tied to improved performance and positive project outcomes. The hypothesis is quite obvious: the best qualified team with the most experience should be able to perform better than a lesser qualified team with less experience. This hypothesis has been tested by a number of research efforts using large datasets and multiple statistical analyses. However, implementing qualifications-driven selection is not always simple given that a contractor's qualifications are more difficult to quantify compared to simply selecting the lowest bidder. Therefore, it is critical to implement a clear and structured selection process that is transparent and fair to all participants. This chapter was developed, based on extensive research and practical experience, as a guide to help design such fair and transparent qualifications-driven procurement processes for design and construction projects. Project owners can benefit by learning how to design or improve their existing procurement process, while engineers and contractors can learn what to expect when competing in a qualifications-driven selection for their next project.

## Background and objective

The Alternative Project Delivery Method (APDM) refers to delivery methods that invite contractor engagement before the design is complete, where construction may start before the design is fully completed and where the project team selection may not solely rely on the lowest cost. APDMs include Construction Management-at-Risk (CMAR), Design–Build (DB), Job Order Contracting (JOC) or Indefinite Delivery/Indefinite Quantity (IDIQ), Integrated Project Delivery (IPD), and others.

Procurement of APDM design and construction projects can vary significantly, but some typical steps exist in most projects. In the typical public works project, a public owner will initiate and execute a series of procurement processes to select separate members of a project team and assemble a team with various skills and disciplines to deliver the required services, such as the design professional, project manager, or general contractor. The owner then assembles these individual participants into a project team to deliver a fully functional facility that meets the owner's needs. The team typically exists only through the design and construction phases of the project with some overlap into the operations phase for warranty services.

There are numerous delivery methods that can be employed in the design and construction of facilities. Procurement activities may vary due to the governing laws and regulations that each project must operate within. As a result, the procurement processes described below will be general in nature. Before applying these procedures, each participant should consult with their own legal counsel to ensure compliance with applicable legal requirements.

Publications exist and provide rich information on the topic of qualifications-driven selection. These include Gransberg and Touran's (2019) synthesis study on the value, benefits, and limitations of qualifications-based selection (QBS) for airport project delivery. The work focuses on airport projects and included a compiled bibliography on the topic. Alleman *et al.* (2017) compared qualifications-based selection and best-value procurement for construction manager–general contractor (CM/GC) highway construction. Shelton (2018) also developed a report on QBS that discusses it as the best practice for architecture, engineering, and construction management/general contractor procurement in Canada. Most recently, Adamtey (2020) published an empirical study to assist owners in selecting the right procurement method for design–build projects. Several research efforts using large datasets and multiple statistical analyses have tested the research hypothesis that the best qualified team with the most experience should be able to perform better than a lesser qualified team with less experience (e.g., Molenaar and Franz, 2018; El Asmar, 2012). Evidently, each of the available publications is focused on one particular scope like a particular type of project (e.g., airports or highways), a specific geographical location (e.g., Canada), or one distinct delivery method (e.g., DB or CMAR). One identified gap in the literature is guidance for project owners on how to design a robust qualifications-driven procurement process.

This chapter aims to fill this gap, summarizing the knowledge on the topic of qualifications-driven procurement and providing guidance resulting from both research findings and extensive practical experience. There is a difference between a pure QBS, which does not include a price component, and a best-value approach, which includes both qualifications and price. This chapter covers both types, to guide any selection method that includes qualifications (either partially or wholly). The chapter discusses how to develop a successful request for qualification (RFQ) that meets typical government requirements. It lays out typical RFQ contents in terms of administrative requirements and evaluation criteria that can be used effectively. It overviews the RFQ and request for proposal (RFP) steps and delineates how this process can inform the procurement decision and final award. Project owners can benefit by learning how to design a procurement process or improve their existing process, while engineers and contractors can learn what to expect when competing in a qualifications-driven selection for their next project.

## Procurement process

The procurement phase for APDM projects can be divided into clearly distinct linear steps. Not all steps are used in every owner/agency's process, but procurement processes almost always start with advertising the project, oftentimes followed by prequalification of teams or RFQ to short-list the most qualified teams for the project. Some procurement processes are known as "one-step" and rely solely on qualifications, skipping the next proposal-focused step and awarding the project based on the competitors' statements of qualifications (SOQ). Some one-step processes consist of just a proposal step, without SOQs. In other instances, sometimes termed "two-step", qualifications-based selections are only used to shortlist a small number of three to five teams that will then enter a more detailed RFP competition, followed by the final selection of the winning proposal. Depending on the agency and contract choices, some processes allow for negotiations at this juncture, to set the stage for the final agreement and award. Regardless of whether the competing teams' qualifications are used as the sole basis for the selection, as part of the criteria, or as the first of two steps, fairness is very important and oftentimes price is viewed as one element that helps ensure a fair selection. For example, at the time of this writing, federal regulations still require cost to be a component of the selection if federal funds are used on a project. Agencies and project owners must be careful to lay out a clear and transparent process that ensures a fair competition and the selection of the most qualified team. The remainder of this section discusses the qualifications-driven procurement process in detail.

A commonly used procurement process is generally described as a "two-phase" process in which the first phase consists of short listing (based on qualifications submitted in response to an RFQ) and the second phase consists of the submission of price and technical proposals in response to an RFP. This two-phase process allows the owner to conduct a "best-value" selection process in which proposals contain both price and qualitative components, and the award is based upon a combination of price and qualitative considerations.

Gransberg and Ellicott (1997) define best value as a procurement that focuses on selecting the contractor with the offer most advantageous to the government, when price and other factors are considered. "Best-value procurements allow government contracting services to evaluate offers on the basis of total procurement costs, construction quality issues, completion dates, additional features and technical innovations. More importantly, they force the early development of detailed project and procurement plans and create solicitations that contain accurate source-selection criteria. This combination of early planning and quality-based contracting yields significant benefits in construction timeliness, cost containment and customer satisfaction."

The process described in this chapter discusses a two-phase process with the issuance of a RFQ followed by a RFP. Issuance of the RFQ and the RFP are the two major phases in the selection process, but other steps are necessary to complete the procurement. The complete list of steps includes:

1. Assess functional needs and identify facility requirements.
2. Prepare selection panel voting members and technical support staff.
3. Prepare and issue RFQs.
4. Receive and evaluate SOQs.
5. Prepare and issue RFPs.
6. Receive and evaluate technical proposals.
7. Conduct interviews and perform due diligence.
8. Evaluate price proposals, finalize selection, and announce results.
9. Negotiate contract.

This is not an exhaustive list and there will be additional sub-elements to each of these activities. Each of these nine steps will be described in greater detail in the remainder of this chapter.

Another common procurement process is called "one-step", in which the selection is based on only one of the RFQ or RFP. One-step processes that are based on an RFQ and consist mostly of qualifications are more successful at attracting high-performing contractors compared to one-step processes that are based on a full RFP and require significant design and a project price competition as the criteria for the selection (Ramsey *et al.*, 2016). In any case, the majority of the nine steps presented in this chapter also apply to one-step processes.

## *Assess functional needs and identify requirements*

The first activity the owner must undertake is to develop a comprehensive, detailed listing of the functional needs of the facility being considered for the project to have any possibility of success. The owner must have a clear objective of what he or she wants to accomplish through the APDM process and to be able to communicate that objective clearly to the industry partner that will be expected to deliver that project. The importance of this task cannot be overemphasized. Many unsuccessful projects have resulted from poor project definition.

Some of the questions the owner should answer in preparing the project assessment include the following: What activities will be performed in the facility? How large does the facility need to be? (depending on the facility type, size can be specified in terms of useable space, say, if the project is a convention center, number and size of rooms for an educational facility, water output for a treatment plant, and so on) Why is the facility needed? Who will use the facility and how often will it be used? What level of quality will the owner expect in the completed facility? When is the facility needed? What might be the possible outcomes if the project is not completed on time? What is the owner's budget for the project and when will the funds be available?

During this step, the owner must also develop an accurate budget or source of funding for the project. If the project will be funded through a tax levy or through user fees generated by the project, the owner must carefully evaluate those cash flows to ensure the project is scaled appropriately. Along with preparing the budget, the owner must prepare a schedule for the delivery of the project. Service providers who will submit proposals for the project will need to understand the owner's schedule in order to price their services effectively.

If the owner lacks APDM experience or staff to prepare the project definition, budget, and schedule, he or she needs to hire an experienced consultant to provide support in performing this task. This information is vitally important in preparing an effective RFQ and RFP.

## *Prepare selection panel voting members and technical support staff*

Early in the procurement process, the owner must assemble a selection panel that will score proposals received in response to both the RFQ and the RFP. In general, the panel should consist of qualified members who are familiar with the project requirements and have the knowledge and experience to properly evaluate the submittals they will be receiving. For public works projects, the specific composition of the selection panel may be dictated by local laws and regulations. For example, the State of Arizona's public procurement laws set a limit of seven members for panels in the selection processes for DB, CMAR, and JOC project procurement. The law further allows up to two additional members if the project involves multiple public agencies. Arizona laws also require that one of the panel members be a professional registrant, such as a registered professional engineer or architect, and that one member be a senior management employee of a licensed contractor.

Selection panel members need to receive training on their duties and responsibilities. At a minimum, this training should acquaint the panel members with the delivery method, the evaluation criteria, and the scoring methodology that will be used in the process. Panel members should be instructed to score all submittals completely and consistently. They should be directed to provide comments and feedback that may be shared with the unsuccessful proposers that will help them be more competitive in the future. Comments should also be factual, objective, and relevant to the scoring criteria. Scoring criteria will be discussed later in this chapter.

Panel members do not need to be subject matter experts to participate in an APDM procurement process, unless local/applicable jurisdictions dictate otherwise. To assist panel members in understanding and evaluating the more complex, technical aspects of some projects, project owners may also establish a Technical Review Committee (TRC). This TRC will include subject matter experts, either from the owner's staff or outside consultants, who will advise the voting panel members on the technical aspects of the submittals. For example, most procurement processes will ask for detailed financial information from the proposing firms to evaluate their fiscal viability and strength. For very large projects that extend several years, a financial analysis expert may be needed as a member of the TRC to interpret the financial reports and advise the selection panel members on the relative importance of the information received. Other technical experts may be added to the TRC as needed.

Panel members can also serve as a resource for the owner to review and comment on the procurement documents. Since the panel members will be evaluating SOQs and proposals based on the criteria in the RFQ and the RFP, there is logic to having them participate in the process of preparing those criteria.

## *Prepare and issue request for qualifications*

Based on information generated in Step 1 described above, the owner will now prepare and issue the RFQ to the interested parties. At a minimum, the RFQ should contain the following information:

- Scope of services.
- Draft budget and schedule.
- Overall selection process details, including the number of shortlisted firms.
- RFQ evaluation plan, including selection criteria (and eventually criteria weights).
- Submittal requirements, form and format, timing, location, and so on.

The functional assessment for the facility prepared in Step 1 serves as the basis for the RFQ. The owner must describe its needs thoroughly through the RFQ so that proposers can craft their responses appropriately. In preparing their SOQs, proposers must present relevant information on successful projects that they have completed and convince the selection panel that those previous successful projects make their team uniquely qualified for the current project. To be the most compelling, proposers need to provide comparable projects. For example, the design and construction of a very successful hotel project in Beirut does not necessarily ensure that a proposer is the best qualified team to design, construct, operate, and maintain a wastewater treatment plant in London. In order for a selection panel to make this evaluation of comparability, the RFQ needs to have a thorough description of the project and a complete listing of the services that the owner is expecting to be provided.

The amount of information to be contained in the "Scope of Services" section of the RFQ can vary widely depending on the project parameters. The owner must exercise judgment in crafting the scope of services to ensure that interested parties fully understand the services required in order to prepare an SOQ that is responsive to the owner's needs.

After describing the project and the services, the owner needs to prepare an evaluation plan that reflects what criteria are most important in the selection. An evaluation plan includes a listing of the evaluation criteria and the relative weights of those criteria. The owner will need to consider the following questions: Are the aesthetics of the completed project more important than the functional performance? Is meeting the owner's budget more important than the project schedule? The owner demonstrates the importance of the selection criteria by assigning a heavier weight to those items that are more important and a lesser weight to those that are less important.

The RFQ should also thoroughly describe all the steps that will be followed during the entire selection process. If the owner desires to interview the teams in Phase 1 (RFQ), Phase 2 (RFP), both phases, or neither phase, that information should be included in the RFQ. The manner in which the owner intends to use the results of the two evaluation phases should also be addressed. For example, are scores from Phase 1 added to the scores from Phase 2 and the sum used to determine the best qualified team? Or are the scores from Phase 1 only used to determine the short-listed firms, while the final recommendation will be solely based on scores from Phase 2? Are interviews, if conducted, scored separately or are they used as additional information by selection panel members in scoring the SOQs or proposals? Those are just few of the options an owner may use during the selection process. Whatever process will be used must be described in the RFQ so that interested parties can understand how they will be evaluated and can make an informed decision on whether or not they want to participate.

Submittal requirements in terms of numbers of SOQs, limitations on the total number of pages, and date and time of submission must be included in the RFQ along with a clear statement of the consequences if the submittal requirements are not followed precisely.

All the above information should be included in the RFQ. Before issuing the RFQ, it is advisable that the document be reviewed for clarity and consistency. Since the selection panel described earlier in Step 2 will have to interpret and apply the selection process described in the RFQ, they should have the opportunity to review and comment on the RFQ before it is finalized and issued. It may also be advisable to conduct a legal review of the RFQ by in-house or outside legal counsel for compliance with applicable procurement codes and regulations (examples of compliance issues include meeting a minimum percentage threshold for the cost component in some states or specific shortlisting requirements) This review may be necessary for unique projects or for projects by an owner using a particular delivery method for the first time. Legal reviews may be expensive, but they may avoid costly mistakes later in the process or avoid a protest of the selection results.

Depending on the procurement rules, there may be a statutory requirement for advertising a project before the issuance of an RFQ. Even if public advertising is not required, it is good practice to issue a press release to generate interest in the project. Depending on the scope and scale of the project, it may be necessary for proposers to form joint ventures or to team up in other less structured ways, in order to pursue the project. This advanced advertising will notify the industry of the owner's intent and allow time for teams to assemble. More targeted advertisements may be necessary to reach a particular sector of the industry that may not be attuned to an owner's normal procurement solicitations.

Any advertisements for the project need to state how the formal RFQ will be issued and distributed. Will the RFQ be issued in hard copy or electronically? Are proposers required to pick up copies or the RFQ at the owner's offices or will it be distributed via mail? Can prospective proposers download the document from a website or will it be sent out as an e-mail attachment?

Regardless of the process for distribution of the RFQ, the owner should maintain a listing of all proposers who have received a copy of the RFQ in the event that corrections or addenda need to be issued. Without this "plan holders list", the owner has no way of knowing who has received the RFQ and no way of disseminating changes or corrections. Without this ability, some proposals may be considered non-responsive if they don't comply with subsequent changes in the RFQ requirements. If made public, this plan holders list will allow interested parties to see who else may have received a copy of the RFQ to help in identifying possible partners or to assess the level of competition.

Another good practice is for the project owner to conduct a presubmittal conference. The presubmittal conference allows the owner to orally present the material contained within the RFQ and answer any questions that the proposers may have. The presubmittal conference also serves to evaluate the level of understanding of the RFQ and selection process and may lead to the issuance of addenda if errors or inconsistencies are identified. The presubmittal conference is typically scheduled at least two weeks after issuance of the RFQ so that interested parties have time to review the document ahead of time and prepare any questions they may have. The conference should also be scheduled at least two weeks before SOQs are due, so proposers have time to revise or amend their SOQs based on information that they have received. It is advisable that the information presented at the presubmittal conference be limited to the information contained in the RFQ. Some owners require mandatory attendance of presubmittal conferences, but that has not been a widely adopted standard practice.

The presubmittal conference also serves as an opportunity for proposers to identify partners and evaluate the competition. The presubmittal conference allows proposers to meet and interact with the owner's project team, which can help to instill confidence in proposers that the selection process will be conducted fairly and impartially. By hearing and observing the owner team in presenting information about the project, proposers can determine if they would like to pursue work with the owner — in effect the presubmittal conference may serve as an interview of the owner's team.

### *Receive and evaluate statements of qualifications*

Upon receipt of SOQs, the owner's staff must evaluate them for responsiveness by applying the requirements identified in the RFQ. Non-responsive SOQs should not be forwarded to the selection panel for consideration. A couple of initial questions for the owner's staff to answer are as follows: Did the proposers submit the necessary number of copies? Do the SOQs comply with any page number limitations? If any submittals are judged to be non-responsive, they are immediately rejected by the owner's staff and returned to the proposer.

Responsiveness criteria must be applied strictly and uniformly, even if the deficiency might be very slight. If the owner makes an exception for one proposer, he or she has then disadvantaged every other proposer and created an uneven playing field for the selection process. Another aspect to consider is that if a proposer cannot comply with the RFQ requirements, which are not overly complicated, are they going to be able to comply with the much more detailed requirements of a contract? If the proposer's attention to detail is lacking in assembling the SOQ, are they going to provide attention to detail on the project itself, if selected?

Responsive SOQs are provided to the selection panel members for consideration along with scoring sheets. Providing scoring sheets to the panel members is a good practice to help the members score consistently. If a proposer chooses to protest the selection results, the owner can defend the selection process more effectively if all the panel members have used similar scoring sheets. Panel members should be given adequate time to thoroughly review and score all SOQs before meeting again to select the finalists. The amount of time needed to review and score the SOQs will depend upon the number of SOQs received and the number of pages allowed for each SOQ.

It is important that all panel members review and score every SOQ thoroughly. Some panel members may have a tendency to sort through the SOQs and only score the most competitive ones completely and ignore other SOQs from the lesser qualified teams. However, good owners owe honest and complete feedback to every single proposer. The lesser qualified teams will never be able to improve and become more competitive if they do not receive feedback on areas where they might be lacking. It is in the owner's best interest to build a strong competitive community of contractors, and such feedback will help achieve that goal in the long term.

After an appropriate amount of time for individual scoring, the selection panel members should meet for a shortlisting meeting. This meeting may take various forms, but the ultimate goal is to arrive at a consensus on the best qualified firms to continue to Phase 2 of the selection process in the form of the RFPs. The number of shortlisted firms may vary based on local procurement rules.

The shortlisting meeting should begin with all the panel members providing their scores for each SOQ to serve as a starting point for further discussion. If there is consistency and a clear consensus of the top teams, there may be little need for further discussion. If there are big differences in scores among panel members, more discussion is required. This discussion may lead to some panel members changing their scores, which is appropriate and acceptable. The value of a diverse, well-qualified selection panel is that the various members bring their own experiences to bear in the process. With effective facilitation, every panel member is allowed to express their perspectives, resulting in a better outcome. Although the results may not be unanimous, ultimately the panel must arrive at a consensus on the requisite number of shortlisted firms.

Individual and group scores need to be recorded and maintained for debriefing purposes and for the possibility of a protest. Again, procurement rules may vary, for example, in public works projects in many jurisdictions, it is required to maintain these scores and they become public records. The identity of individual panel members may be protected, but the scores and panel member comments may not be protected.

The results of the shortlisting meeting need to be posted publicly as soon as possible. Shortlisted firms are notified, and instructions are provided to them regarding the next step in the selection process, which is typically the issuance of a RFP. Unsuccessful firms are also notified and offered the opportunity of a debriefing on the shortlisting process.

## *Prepare and issue request for proposals*

Drafting the RFP is the next step for the project owner. Considering the applicable procurement laws and policies provides a basis for the remainder of the process. Then, given that the RFP evaluation typically consists of both a technical proposal and a price proposal, the RFP development process often starts with discussing and determining the relative weights of these two components of the proposals. Once this has been determined, the owner should dig deeper to set the criteria that make up each of the technical proposal score and the price proposal score. This step is similar to the RFQ preparation discussed earlier, where the owner develops and weighs individual criteria that are important for the project. Based on the authors' analysis of procurement criteria for more than 50 APDM projects, examples of criteria commonly used in both SOQs and RFPs include the following:

(a) Firm experience and qualifications, including experience with the delivery method used on this project.

(b) Key personnel experience and qualifications, including a requirement to submit curriculum vitae and comparable projects.
(c) Project understanding and approach, including expected challenges and how the proposer plans to handle those.
(d) Past performance, including on projects of similar size, type, location, owner, complexity, as well as references from previous clients.
(e) Design approach, including the ability to meet specific design requirements.
(f) Ability to meet the project schedule and budget.
(g) Price component, fees, or cost.
(h) Legal and financial status of the proposing firm(s).
(i) Safety record of the proposing firm(s).
(j) Other criteria that the owner deems critical to this project's success. Examples include alternative technical concepts if applicable, operations and maintenance if the project is a public–private partnership, and so on.

Establishing submittal requirements, including a procurement schedule, is needed to set the expectations with the shortlisted proposers. Good RFPs also provide directions on how to submit questions to the owner. Once these items have all been addressed and documented in the RFP, the owner issues the draft RFP to all the shortlisted firms. It is also a good practice to attach a draft contract so that the proposers can better understand the expectations from a contractual perspective.

After the RFP has been issued, the owner can start planning to conduct a pre-proposal meeting (sometimes called a pre-proposal conference). This is a meeting with the shortlisted firms, where the owner discusses the process described in the draft RFP, answers questions, and provides oral clarifications. It is typical for the owner to gather comments from the pre-proposal meeting, and he or she can issue a revised RFP if deemed necessary. At this stage, the final RFP is issued, and the competing teams start crafting their proposals.

### *Receive and evaluate technical proposals*

The order of the next steps is important. First, similar to the SOQ stage, when the proposals are received, the owner should check proposals for responsiveness. The responsive proposals are then provided to the panel members for their evaluation. Then, the evaluators must completely score the technical proposals before opening the corresponding price proposals, to avoid any possible biases based on the price components. This specifically applies to RFPs that require separate technical and price proposals, oftentimes submitted in different envelopes. Note that some RFPs do not specifically separate the two components (technical proposal and price) and instead include both in the same submittal. In other cases, evaluation panels could be separated into two sub-panels (one for the technical aspects and the second for commercial aspects), although this practice is not as common.

## *Conduct interviews and perform due diligence*

Holding interviews of the competing teams is the next step in the process, to allow teams to communicate their proposals orally. It is often important to evaluate team chemistry since this may be a long-term relationship over the life of the project. Interview settings also allow panel members to ask in-depth, follow-on questions, to clarify key aspects of the proposal. Often, an important benefit of interviews is judging the capabilities of individual team members, to decide if the proposing team is really capable of delivering a successful project. Conversely, the owner may be able to sort out those teams that can write a slick proposal but may not be able to deliver the required services. Therefore, interviews are critical to identifying the best project team as part of a successful procurement process.

The owner also needs to do their due diligence. Due diligence strategies include checking the references of shortlisted firms to be interviewed, developing a standard list of questions to ask each reference (ask the project manager to add and edit questions as appropriate for this project), determining who to contact in the reference list (typically the proposers will list various owners/clients, some of whom may have a professional relationship with the project owner), and preparing a standard form to record results of the due diligence. When ready, the owner should initiate contact with the references and ask his or her questions, prepare the results as a report to the selection panel, and report findings to the selection panel (typically on the day of the interviews, to inform the final scoring).

Based on the results of the interviews and the due diligence, panel members are asked to re-score the technical proposals and make any needed adjustments to their initial scores. At this time, panel members meet, share their scores, discuss any differences, and arrive at a consensus for aggregate scores, which will now be considered the final scores of the technical proposals.

## *Evaluate price proposals and finalize selection*

It is considered a good practice to require price proposals to be submitted separately in a sealed envelope. It is the owner's responsibility to maintain these price proposals in a secure manner (i.e., locked in a safe) until the technical proposal evaluation has been fully completed. At that time, the price proposals should be opened in front of the entire selection panel. Then, the scoring method and weighting is applied as clearly described in the RFP to determine the price score for each proposer. The owner and panel are now ready to combine the technical criteria score and price score into a final score for each proposer. Examples include dividing the technical score by the price, or vice versa, or treating the price component as one of the criteria with a corresponding weight and calculating a weighted average for each proposer.

Calculating the final scores allows the panel to recommend the best qualified proposal based on all the previous steps of the process. Typically, the formulas are such that the highest score wins. The selection panel forwards their recommendation to the appropriate body for final approval. For example, for a city/municipality project, the recommendation would be forwarded to the mayor and city council for approval. Once approval is received, the owner announces the selection results and debriefs the participants. In some cases, non-winning bidders are partially compensated for their efforts using a monetary stipend.

## *Negotiate contract*

Depending on the agency and contract choices, some processes allow for negotiations at this juncture, to set the stage for the final agreement and

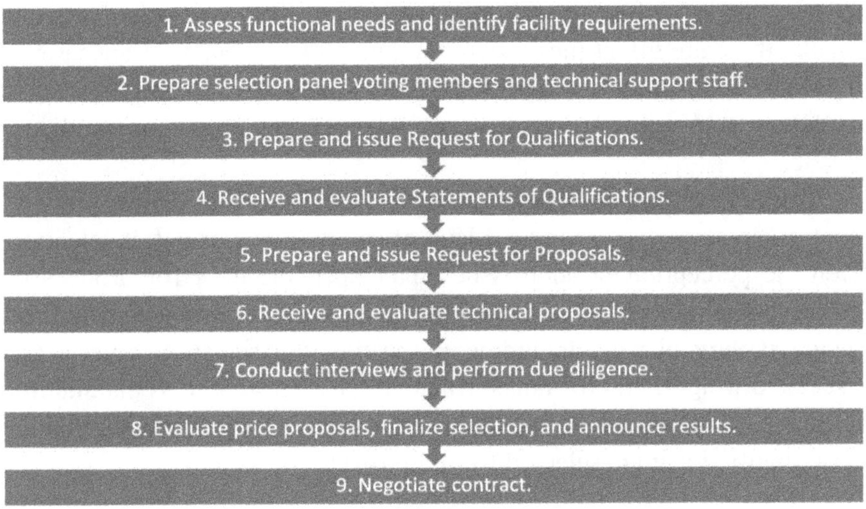


**Figure 1:** Qualifications-driven procurement process.

award. Once the selection results are announced, the owner enters into negotiations with the best qualified proposer. Negotiations continue until reaching an acceptable agreement. If the owner is unable to reach an agreement, he or she can initiate negotiations with the second-best qualified proposer, but in that case the owner cannot return to the first proposer after ceasing the negotiations with them. Although this very rarely happens, if unable to reach an agreement with any of the shortlisted firms, the owner can terminate the process. He or she can initiate a new process, change the parameters to achieve a different result, and even explore other delivery methods and approaches. Again, this is an unlikely outcome if the process was managed appropriately, and expectations were clarified upfront in the RFQ and RFP phases.

To conclude, Figure 1 summarizes the nine steps that make up an APDM qualifications-driven procurement process.

## Administrative considerations

If the procurement is for a public project, agencies often have a requirement to retain procurement records for a period of time. The owner will need to handle public records requests, even after project completion. Finally, there are oftentimes specific requirements about the transparency of the selection process for publicly funded projects, but transparency is actually recommended regardless of whether the project is public or private.

Common pitfalls to watch out for in a qualifications-driven procurement process revolve around the lack of transparency or unfair selection. Even the perception of such issues may open the door for protests by unsuccessful bidders. Documenting the procurement process upfront, following it very closely, including publishing the selection criteria and associated weights and formulas, where applicable, as well as maintaining fairness to all participants, can help avoid protests while ensuring that the most qualified team is selected for the project.

## Conclusions

This chapter discussed qualifications-driven procurement for APDM projects. The authors broke down the procurement into nine steps occurring in linear fashion. Not all steps discussed in this chapter are used in every owner's process, but robust procurement processes often include a combination of these steps needed to select the most qualified team for a given project. Regardless of whether the competing teams' qualifications are used as the sole basis for the selection or as part of the criteria, QBS has been tied to improved performance and positive project outcomes. The highest qualified team, with the most experience, typically performs better than a lesser qualified team with less experience. However, selecting a project team based on qualifications is not always simple given that qualifications are more difficult to quantify compared to simply selecting the lowest bidder. Moreover, fairness and transparency are very important, and oftentimes price is viewed as one element that helps ensure a just selection. Therefore, agencies and project owners must be careful to present a clear and transparent process that ensures a fair competition and the selection of the most qualified team. This chapter provided guidance resulting from both research findings and extensive practical experience, discussing how to develop a successful RFQ laying out typical RFQ contents in terms of administrative requirements and evaluation criteria that can be used effectively, overviewing the RFQ and RFP steps, and delineating how this process can inform the procurement decision and final award.

# About the Authors

**Mounir El Asmar** is an Associate Professor at Arizona State University's Del E. Webb School of Construction. He also is a Senior Sustainability Scientist with the Global Institute of Sustainability and Co-Director of the National Center of Excellence on SMART Innovations. Dr. El Asmar has presented his work on innovative project delivery at more than 100 national and international venues. His 100+ technical articles are published in several archival journals and conference proceedings, and his scholarship has received more than 30 recognitions and awards, including the ASCE Thomas Fitch Rowland Prize, CII Distinguished Professor Award, and DBIA Distinguished Leadership Award.

**Wylie Bearup** served as the Professor of Practice and Beaver-Ames Heavy Civil Engineering endowed chair at ASU. Dr. Bearup worked for the City of Phoenix as the City Engineer and Street Transportation Director. While with the City, he pioneered the use of APDM, creating processes and procedures to take full advantage of the new methods. Under his direction, the City successfully completed more than 400 projects using Construction-Manager-at-Risk, almost 100 Design–Build projects, and approximately 5,000 projects using Job Order Contracting with an aggregate value greater than $6 billion. Dr. Bearup was selected as one of the Top Ten Public Works Leaders in the country by the American Public Works Association.

# Chapter 8

# Addressing the Fallacy in Lump-Sum Best-Value Procurement for Design–Build Highway Projects

Maria Calahorra-Jimenez*, Cristina Torres-Machi† and Keith R. Molenaar†

**California State University, Fresno CA, USA*
*†University of Colorado Boulder, Boulder CO, USA*

## Abstract

Lump-sum best-value procurements of design–build projects enable the evaluation of technical factors in addition to price when selecting design-builders. This chapter examines procurement results from 347 highway projects in the United States over the last two decades. This analysis shows that price was the deciding factor in more than 80% of the selections. In fact, the weighting and scoring of technical proposals made it unlikely that a superior technical proposal would be awarded the project. This chapter explores how historical trends on scores and weights can be used to understand the current bias of best-value procurement toward lowest price. An illustrative example analyzing the impact of weighting and scoring on procurement outcomes is presented. This analysis leads to somewhat counterintuitive results highlighting the importance of scoring ranges in best-value selection. Indeed, the range used to evaluate technical proposals can have a greater impact on the selection than changes in weights. From these results, practical recommendations are derived to help highway agencies and other owners better understand the impact of scoring on best-value selection and how to adjust weights based on expected bid dispersion. These recommendations will guide agencies in authoring best-value procurement plans and determining weights and ranges of scores to promote best-value selection.

## Introduction

Public agencies have traditionally used the design–bid–build (DBB) project delivery system with the low-bid procurement method to develop projects. Design–bid–build is a widely accepted project delivery system in which the design and construction phases are procured independently, resulting in clearly established roles. In DBB, the construction phase is often procured on low-bid, providing an open competition based on cost (Pakkala, 2002; Scott *et al.*, 2006). The traditional DBB approach, however, has been found to generate fragmentation and limit innovation, resulting in higher odds of overruns and disputes between parties (Pakkala, 2002; The American Institute of Architects, 2007).

To address some of the disadvantages inherent in DBB, more collaborative contracting strategies, such as design–build, were developed in the past century. In design–build, public agencies procure both the design and construction in a single contract, enabling more flexibility and collaboration in the development of the design and construction. Some of the advantages of design–build include potential cost efficiencies, opportunities for value engineering, flexibility in risk allocation, reduced litigation throughout the project, and enhanced schedule performance (FHWA, 2006; Hale *et al.*, 2009; Shrestha *et al.*, 2012).

Design–build studies have found that design–build can outperform DBB in different project types, including buildings (Shrestha and Fernane, 2017), industrial and mechanical projects (Franz *et al.*, 2020; Riley *et al.*, 2005), and highway projects (Asmar *et al.*, 2020; Tran *et al.*, 2018). In the United States (US), design–build has seen continuous growth as a delivery method for highway projects — between 2013 and 2017, design–build was used to procure 35% of the highway projects, with this figure projected to grow up to 43% between 2018 and 2021 (FMI, 2018).

Design–build uses a procurement approach that differs from the traditional practices of hiring designers and constructors independently and based on different evaluation criteria. For example, in DBB, the lowest price based on a 100% finished design was the basis for the contractor's selection. By contrast, cost could not be the basis for designers selection, based on the Brooks Act Legislation (Gransberg *et al.*, 2006). The use of design–build instead of DBB shifts this procurement paradigm by not only considering cost but also non-cost criteria to evaluate and select design-builders under one unique contract. This fact led to substantial changes on how procurement had been developed so far.

In the US, the Florida Department of Transportation (FDOT) started the first documented design–build program in 1987 (Ellis *et al.*, 1991; FHWA, 2006). In the late 90s, the Federal Highway Administration (FHWA) promoted the Special Experimental Project Number 14 (SEP-14), aimed at evaluating non-traditional contracting practices and their impact on cost and schedule performance. In 2002, based on SEP-14 results, design–build was permitted by the FHWA through the issuance of the Design-Build Contracting Final Rule (US Federal Government, 2002). The 2002 Rule suggests a two-phase procurement for design–build projects: (1) shortlisting qualified teams based on the Request for Qualifications (RFQ) submitted by the proposers and (2) assessing technical and price proposals (i.e., best-value procurement) received in response to a Request For Proposal (RFP).

Outside the US, by 2002, a study conducted by the Finish Road Enterprise analyzing delivery methods from Australia, Canada, England, Finland, New Zealand, Sweden, and the USA also indicated that the procurement should be different "when selecting a contractor to complete the project and provide services, products, actual construction and other requirements". The report indicated that owners were willing to change the low-bid approach in order to include quality and technical criteria in the selection process (Pakkala, 2002).

Thus, in contrast to the traditional DBB, in which the procurement is driven solely by cost (i.e., low-bid procurement), design–build often considers lump-sum best-value procurement. This procurement method (referred hereafter to as best-value) seeks to award the project to the proposal bringing the most value for the funds invested. Under this premise, best-value procurement considers other factors in addition to cost when evaluating contractors who have placed bids. Ideally, by transferring from low-bid to best-value procurement, agencies would increase the value added to the project for each dollar spent (Abdelrahman *et al.*, 2008).

Despite the foundational best-value premise that procurement considers factors in addition to cost, recent studies have found that the current best-value procurement practice is biased toward low-bid awards. Gaikwad *et al.* (2021) analyzed 305 highway projects procured using best-value in the US and found that 80% of the projects were awarded to the lowest bidder (Gaikwad *et al.*, 2021). These results are aligned with a study conducted by FMI (2018) in which the analysis of 115 design–build highway/street projects concluded that 87% of these projects were awarded to the lowest bidder.

Data shows, therefore, that despite having authorization for best-value procurement in design–build highway projects since 2002, agencies overwhelmingly award projects to the lowest priced proposer. Perhaps, this is not surprising given the culture of only using the DBB project delivery system and low-bid procurement method. This bias toward low-bid neglects the opportunity of best-value procurement to enhance value through the design–build procurement process. Indeed, it incentivizes design-builders to focus on price rather than value.

Several studies have been developed to improve best-value procurement under different approaches. For example, a common concern when using best-value procurement in public projects is that the evaluation of non-cost criteria can be subjective (compared with cost). Previous research has addressed subjectivity and transparency issues in best-value procurement (Abdelrahman *et al.*, 2008; Calahorra-Jimenez *et al.*, 2020; Molenaar and Tran, 2015; Tran *et al.*, 2017). Abdelrahman *et al.* (2008) proposed a model to determine specific non-cost evaluation criteria based on each project's characteristics. The National Cooperative Highway Research Program (NCHRP) developed Synthesis 471 focused explicitly on "practices for developing transparent selection procedures" (Molenaar and Tran, 2015). Tran *et al.* (2017) explored procedures and practices to promote transparency across 46 highway agencies.

Calahorra-Jimenez *et al.* (2020) proposed a structured approach to articulate non-cost evaluation criteria that are comprehensive, direct, unambiguous, and understandable.

This research led to a more objective best-value procurement process. However, few research studies have explored the bias of the current best-value procurement toward low-bid and how the weighting and scoring of proposals may impact the selection.

The exploration of best-value procurement and selection results requires a large set of project data that have not become available until recently when two decades of design–build projects have been awarded. Calahorra-Jimenez *et al.* (2021) began this exploration by using a first-of-a-kind dataset of 882 non-cost scores and 1,158 cost scores from 347 best-value highway projects. Building upon Calahorra-Jimenez *et al.* (2021), this chapter demonstrates trends in scoring that have not been possible to examine before and provides insights into how highway agencies can define best-value procurement to enhance the design-builder opportunity to add value. Since the influence of non-cost criteria in the selection depends on how the evaluation criteria are articulated, weighted, and scored (Calahorra-Jimenez *et al.*, 2020, 2021), this research explores how weights and scoring trends might influence the lowest bidder's probability of winning.

The remainder of the chapter is structured in the following manner. First, basic concepts related to best-value procurement are introduced. Then, the data and methods used for the analysis are presented. This data are then used to analyze industry scoring trends and develop an example analyzing the impact of current scoring trends in best-value selection. Finally, a discussion of results, their practical implications, and recommendations to minimize the historic cost bias of best-value selection are provided.

## Best-value criteria, scoring, and award algorithms

Best-value procurement comprises three main components: evaluation criteria, scoring, and award algorithms. Evaluation criteria include cost and other key factors used in the evaluation and selection process.

The scoring refers to the rating system used to quantify the evaluation criteria. Finally, award algorithms are used to combine the evaluation criteria and scores to obtain an award score that will be used to select the best proposer (Scott *et al.*, 2006).

In best-value procurement, non-cost evaluation criteria should be those that, in addition to price, add value to the project, client, or project outcomes (Scott *et al.*, 2006). The non-cost evaluation criteria should "represent the key areas of importance and emphasis to be considered in the source selection decision" and "support meaningful comparison and discrimination between and among competing proposals" (US Federal Government, 2002). Non-cost evaluation criteria vary depending on each project's goals, objectives, and characteristics (Molenaar and Tran, 2015). Some of the common non-cost evaluation criteria are prequalification, design-builders' past performance, experience, and qualifications or relevant personnel, project management plans, and design approach (Anderson and Russell, 2001; Molenaar *et al.*, 2005, 2014; Molenaar and Tran, 2015; Scott *et al.*, 2006).

Scoring represents the value resulting from the assessment of evaluation criteria. According to Scott *et al.* (2006), there are four scoring system categories: "(1) satisficing (more commonly called "Go/No-Go"), (2) modified satisficing, (3) adjectival rating, and (4) direct point". Evaluation criteria and scoring systems should be aligned with the project goals and be able to extract valid and purposeful information to assess the proposals (Calahorra-Jimenez *et al.*, 2020).

Award algorithms synthesize the steps that highway agencies should follow to combine the evaluation criteria's scores and provide an award recommendation. In highway projects, seven different award algorithms are commonly used (Scott *et al.*, 2006). Each of them will offer a different technical cost trade-off in lump-sum best-value awards. According to Molenaar *et al.* (2014), the seven types of award algorithms are as follows:

*Fixed budget/best proposal*: Using this award algorithm, highway agencies establish contract price in the proposal and the non-cost evaluation factors. Then, the proposals are evaluated based on non-cost factors since the price is fixed.

*Adjusted bid*: In this award algorithm, the price proposal is adjusted using the technical score. This adjustment is usually obtained by dividing the project price by the non-cost score. The proposer with the lowest adjusted bid will win the project. The advantage of this formula is its similarity to the low-bid award (lowest score wins). The disadvantage is that it can be difficult for proposers to determine the effective weight of the non-price factors.

*Adjusted score*: By using this award algorithm, highway agencies assess the technical proposal first and assign a technical score. In the second step of the process, the price is opened and the score is adjusted. The highest scored proposal is awarded the project.

*Weighted criteria*: In this award algorithm, a weight is assigned to the price and each of the non-cost evaluation factors. The total score is obtained by adding all these values. The highway agency will then select the proposer that gets the highest total score. The advantage of this formula is its transparency related to the weight of each evaluation criterion. Equation (1) is an example of the weighted criteria algorithm, which will be used for illustration later in this chapter.

$$F.S. = Wc * Sc + Wnc * Snc, \quad \text{with } Wc + Wnc = 1 \tag{1}$$

where F.S. is the final score, $Wc$ and $Wnc$ are the weights of cost and non-cost factors, and $Sc$ and $Snc$ are the score of cost and non-cost factors, respectively.

*Meets technical (non-cost) criteria — low-bid*: In this award algorithm, highway agencies evaluate all the non-cost criteria using a predetermined rating system. At the end of this process, only proposals that meet the technical criteria are considered for the next step. The proposal's price component is opened in the next step, and the lowest bidder is the winner.

*Quantitative technical cost trade-off*: In this award algorithm, highway agencies establish a method to score that considers the incremental opportunities of the cost and non-cost benefit. The non-cost score's increment is obtained by identifying the highest non-cost score and dividing it by the next higher score. The determination of the cost score is performed similarly. The proposer with the lowest cost wins the project unless a higher-cost proposal offers justifiable benefits to the project.

*Qualitative technical cost trade-off*: According to Scott *et al.* (2006), the final decision in this method consists of an "evaluation, comparative analysis, and trade-off process that often require a subjective judgment on the part of the selecting official".

## Data collection, methods, and industry trends

The study followed a three-step approach (Figure 1) that considered historical data presented in Calahorra-Jimenez *et al.* (2021). Statistical analysis and distribution fitting were used to characterize scoring trends based on the historical data. Based on the fitted cost and non-cost scores distributions (i.e., industry scoring trends) and an illustrative case study, the probability of the lowest bidder winning in different scenarios of weights was analyzed. The aim of the analysis and discussion is to inform practitioners on how scoring practices influence the probability of the lowest bidder to win.

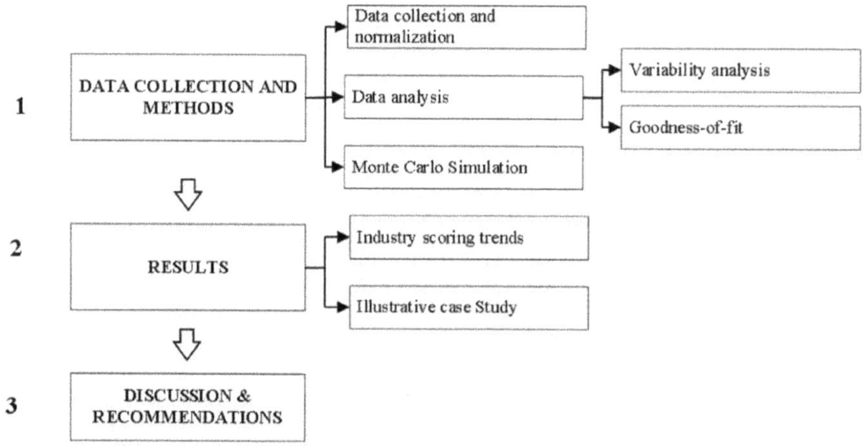


**Figure 1:** Research approach.

## *Data collection and normalization*

The analysis was based on the cost and non-cost database compiled by Calahorra-Jimenez *et al.* (2021). This database comprises cost and non-cost scores from the procurement of 347 design–build best-value highway projects. These projects were procured between 2002 and 2020 by 22 Departments of Transportation across the United States. These data were collected from DOTs' websites and direct requests via email to DOTs' representatives. Two to seven bids were available for each of the projects, with each of these proposals receiving one cost score and one non-cost score. The 347 best-value procurements resulted thus in 1,158 cost scores and 822 non-cost scores. Because some agencies do not make the non-cost components of the results available, the dataset contains more cost points than non-cost scoring points. These additional cost scores were retained to provide a more robust cost trend analysis.

The cost scores in each procurement relate to the evaluation of the economic proposals. To transform the dollar cost proposal into a dimensionless score, the economic bids were normalized based on the proposal that offered the lowest price (Eq. (2)). Usually, highway agencies adopt this practice when using weighted criteria best-value procurement (Molenaar and Tran, 2015), and this is the approach taken in this research to understand and evaluate historic cost scoring trends. As a result of the normalization, the lowest bid is assigned a cost score of 1. The other proposers received a normalized score between 0 and 1 based on how their bid compared to the lowest bid.

$$Sc_i = \frac{Lowest\ bid}{Proposer_i\ bid} \tag{2}$$

where $S_{ci}$ is the normalized cost score on a scale 0–1, for proposer *i; the Lowest bid* is the lowest price bid among all the proposers; the $Proposer_i$ *bid* is the price of proposer *i*.

The non-cost score given to each company is obtained by assessing a set of different technical criteria. Usually, the technical criteria relate to a subset of the following: (1) design, (2) geotechnical investigations, (3) construction, (4) quality, (5) schedule, (6) environmental compliance, (7) capabilities and experience, (8) project understanding, (9) utilities, (10) maintenance and protection of traffic, (11) safety, (12) project management, (13) responsiveness with the request for proposal, (14) right of way, (15) collaboration, and (16) long-term maintenance (Calahorra-Jimenez *et al.*, 2021; Scott *et al.*, 2006).

The technical criteria can be assessed using different scales (e.g., 0–100, 0–1,000, 0–1,200, etc.). To facilitate project comparison, a common zero-to-one scale was used to normalize the technical scoring. This normalization was performed using Eq. (3), which guarantees the proportionality between scales so that the relative distance between scores is the same for the original scale and the normalized scale.

$$Snc_i = \frac{Initial\ nc\ score_i}{Max\ initial\ nc\ score} \tag{3}$$

where $Snc_i$ is the normalized non-cost score for proposer *i*; $Initial\ nc\ score_i$ is the non-cost score of proposers *i* based on the initial procurement scale; *Max initial nc score* is the maximum value of non-cost scores based on the initial procurement scale.

## *Variability analysis*

The DOTs participating in this study had different levels of expertise in best-value procurement. As a result of this, some DOTs contributed to a larger set of data. For example, Florida contributed to 35% of the whole sample due to the significantly larger number of design–build projects they have completed when compared to the other DOTs (FHWA, 2018). To ensure that the scoring data were homogeneous in terms of geographic distribution, a variability analysis was performed. The variability analysis sought to test statistically significant differences in the scores of Florida's projects and the overall sample. A Mann–Whitney U test was conducted, resulting in a $p$ value of 0.154. There is, therefore, no statistical difference in scoring between Florida and other DOTs and the dataset can be considered as consistent.

## *Goodness-of-fit analysis*

This dataset was then used to define probability distributions describing current trends in cost and non-cost scoring. A set of probabilistic distributions were explored, and goodness-of-fit techniques were applied to measure how accurately these distributions fit the dataset (D'Agostino and Stephens, 1986). In this analysis, the goodness of fit was assessed using Akaike Information Criteria (AIC) and the Bayesian Information Criteria (BIC). Although these metrics do not have physical meaning (Yoe, 2019), they can be used in relative terms to compare different distributions, with the lower parameter indicating a better fit. From this analysis, the probability distributions having the lowest values for both AIC and BIC methods were selected.

### *Monte Carlo simulation*

To understand how technical scoring, bid dispersion, and weights of cost influence the probability of awarding the contract to the lowest bidder, a Monte Carlo simulation was used. Using Monte Carlo simulation, it is possible to analyze a phenomenon using random numbers generated from probability distributions defined by the researcher (Johnattan Mun and Pomering, 2010; Shonkwiler and Mendivil, 2009).

In this research, the metric under analysis is the final score (*F.S.*) resulting from the award algorithm (Eq. (1)). Monte Carlo simulation was used to obtain cost and non-cost scoring ($Sc$ and $Snc$) based on the probability distributions describing historical scoring trends. These results were analyzed considering different weights of cost ($Wc$). This resulted in probabilistic distributions of the final scores (*F.S.*) for different scenarios of weights.

In a given procurement, the company that is awarded the contract is the one with a higher final score (*F.S.*). To analyze the probability of the lowest bidder to win, this research simulates the final score of two firms, one of them being the lowest bidder (i.e., $Sc = 1$) and then calculates the probability of the lowest bidder's final score to be higher than the *F.S.* of the other firm.

## Results

### *Industry scoring trends*

Based on the goodness-of-fit analysis, the beta distribution was found to fit best cost scores. Other distributions (e.g., normal, logistic, triangular)

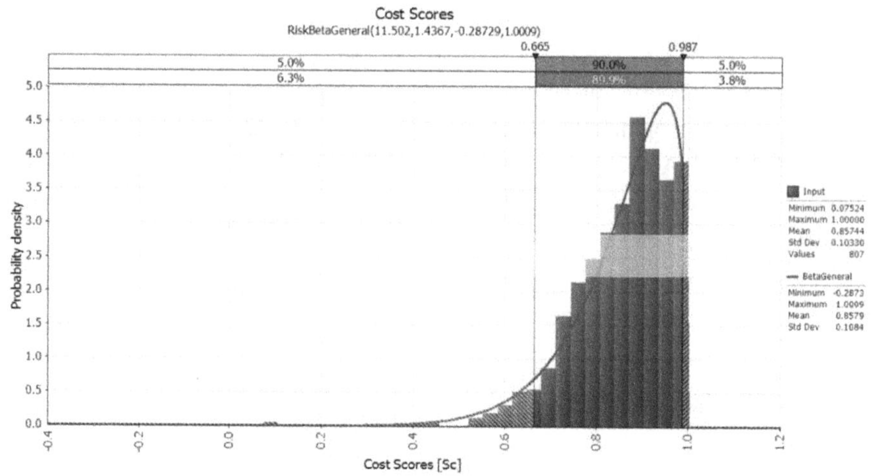


**Figure 2:** Cost scores histogram and PERT probabilistic distribution.

were explored but resulted in higher values of AIC and BIC and were then deemed as worse fits (Calahorra-Jimenez *et al.*, 2021). The beta distribution ranges between 0 and 1 and is commonly used to describe variability over a limited range (Yoe, 2019). It is characterized by four parameters: two shape parameters and two bound parameters (lower and upper bound). Using a beta distribution, cost scores are characterized by a maximum value of 1 and a 95% probability of being between 0.665 and 0.987. The graphical representation of this distribution and the empirical data is shown in Figure 2.

A Gumbel distribution was found to be the best fit for non-cost scores. The Gumbel distribution is often used to model the maximum and minimum values of any set of data (Gumbel, 1955), such as extreme events and construction design elements (Johnatahn Mun, 2002). This distribution is characterized by a location and a scale parameter. As shown in Figure 3, the non-cost scores are skewed toward the end of the evaluation scale (i.e., 90% of the data ranges between 0.718 and 0.96), and the Gumbel distribution resulted in good representation of non-cost scores.

The historical trends of technical and cost scores show two important characteristics. First, both distributions are skewed toward 1. Second, in more than 90% of the cases, both the technical and economic scores are

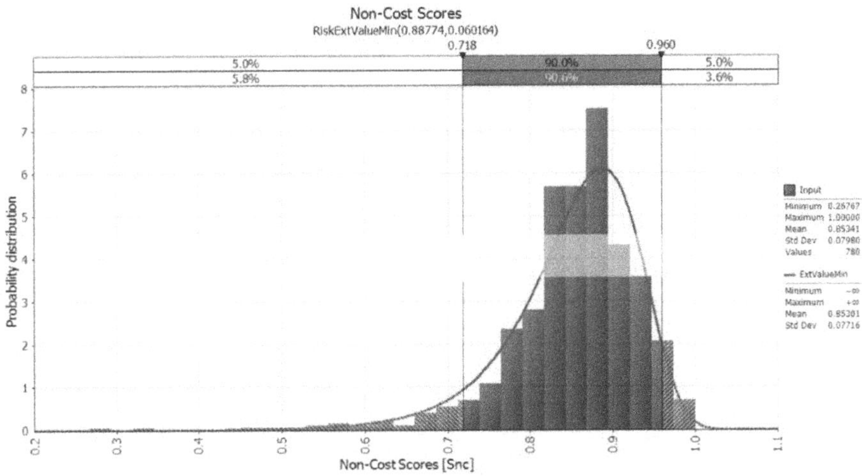


**Figure 3:** Non-cost scores histogram and Gumbel probabilistic distribution.

higher than 0.7 (Figure 4). This results in a narrow range of variation in scoring practices.

The cost score represents the variability of bids in relation to the lowest bid and can therefore be used to evaluate bid dispersion. Multiple factors can drive the dispersion of bids. For example, projects with a clear scope, few risks, and minimal room for innovation in design will result in low-bid dispersion and low variability in bid scores. Non-cost scores, on the other hand, represent the evaluation of technical proposals. Having a narrow variation of technical scores implies a lack of differentiation among the proposers regarding the evaluation criteria used in the assessment of the proposals. This lack of differentiation can stem from a variety of factors such as similarly qualified design–build teams, little room for innovation in design proposals, or the use of a narrow range of scores when evaluating technical proposals. All these factors will result in a narrow distribution of the non-cost scores.

## *Illustrative example*

This section presents an illustrative example analyzing the probability of the lowest bidder winning a design–build best-value procurement with

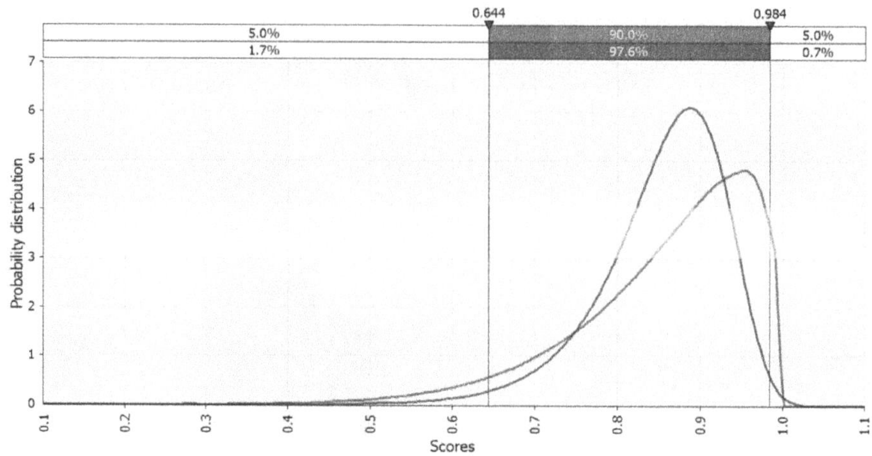


**Figure 4:** Cost (red) and non-scores (blue) probabilistic distributions.

two proposers. This example is intended to illustrate the implications of current scoring practices and how this may result in biased selections toward low-bids.

In this example, company A represents the lowest bidder. The lowest bidder's cost score (*Sc*) is 1, while the non-cost score would be represented by the probability distribution obtained previously. For company B, the cost and non-cost scores will be represented by the respective probability distributions. The final score for each company can be obtained as follows:

$$\text{Company A (lowest bidder): } FS_{\mathrm{A}} = Wc * 1 + Wnc * Snc \tag{4}$$

$$\text{Company B: } FS_{\mathrm{B}} = Wc * Sc + Wnc * Snc \tag{5}$$

The distributions considered for cost and non-cost scores are *Sc* ~ BetaGeneral (11.50, 1.43, –0.28, 1.00) and *Snc* ~ Gumbel (0.88, 0.06).

Monte Carlo simulation was used to obtain the probability of the lowest bidder to win (i.e., $FS_A > FS_B$) for different scenarios of weights of cost (*Wc*). Figure 5 shows that for a weight of cost varying between 30% and 70%, the probability of the lowest bidder winning ranges between 70% and 94%.

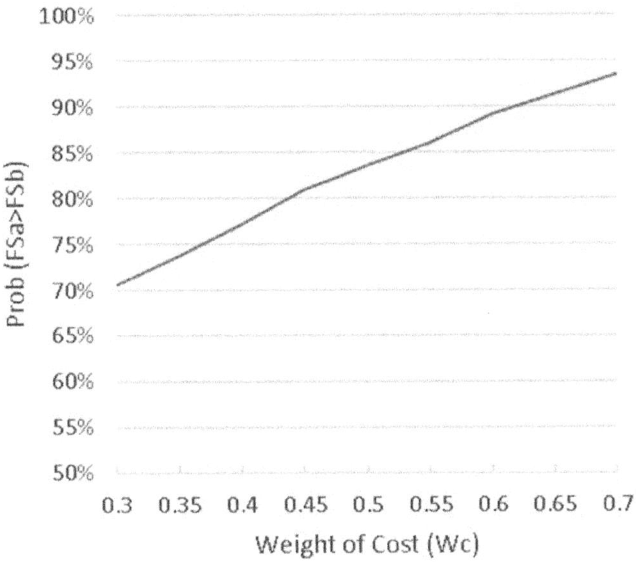


**Figure 5:** Probability of the lowest bidder to win depending on the weight of cost and current scoring practices.

In decision-making, weights are aimed at representing the relative importance of the evaluation criteria and represent the decision-maker's preference toward these criteria. Intuitively, one would expect that assigning equal weights to cost and non-cost factors (i.e., $Wc = 50\%$) would result in 50% chances of the lowest bid to be selected. The results obtained in the simulation, however, show differently. Figure 5 reveals that even when the weight given to the cost component is low (e.g., $Wc = 30\%$), the lowest bidder has very high chances (i.e., 70%) to win the bid. This is probably due to the high skewness and low variability of scoring trends (Figure 4). Figure 5 shows that the current trends in scoring result in an increased probability of the lowest bidder winning and distort the effect of cost and non-cost weights in the award algorithm.

The results of this illustrative example may help explain the reasons why more than 80% of highway projects procured through best-value design–build in the United States were awarded to the lowest price. It also provides a model to explore how changes in scoring approaches and

**Table 1:** Non-cost simulation distributions to explore wider scoring ranges.

| Variable | Distribution |
|---|---|
| *Sc* | Beta General (11.50,1.43, –0.28,1.00)* |
| *Snc* Historic | Gumbel (0.88, 0.06) |
| Model 1 | Gumbel (0.88, 0.10) |
| Model 2 | Gumbel (0.88, 0.15) |

*Note*: *All the distributions are truncated at 0 and 1.

anticipated dispersion in price proposals can change the award outcomes. Specifically, the remainder of this section will explore how changes in scoring can increase the probability of awarding the contract to a proposer who does not submit the lowest price proposal.

To analyze the impact of different non-cost scoring trends on the outcome of a procurement, two alternative Gumbel distributions were explored and compared to the one describing historical trends. These alternative distributions, depicted as Model 1 and Model 2 in Table 1, have the same mean as the historical distribution but higher standard deviations to represent wider scoring ranges.

The distributions used in Models 1 and 2 for non-cost scoring have the same location parameter (i.e., 0.888) as the historical trend but consider a higher scale parameter (i.e., 0.1 and 0.15 vs. 0.060) to represent wider scoring ranges. In Gumbel distributions, as the scale parameter decreases, the distribution becomes taller and narrower. As the scale parameter increases, the distribution becomes shallower. To analyze the effect of flattering the distribution of non-cost scores, Models 1 and 2 were created by increasing the shape parameter to 0.1 and 0.15. The graphical representation of historic Model 1 and Model 2 non-scores distributions is shown in Figure 6.

Based on Eqs. (5) and (6), the probability of the lowest bidder to win (i.e., $FS_A > FS_B$) was simulated under different scenarios of weights for different distributions for the non-cost scores. Figure 7 shows how the probability of the lowest bidder winning varies in each of these cases.

The results show that higher ranges of variation for technical scores decrease the probability of the lowest bidder winning. In other words, the probability of awarding the contract to the lowest bidder is influenced by

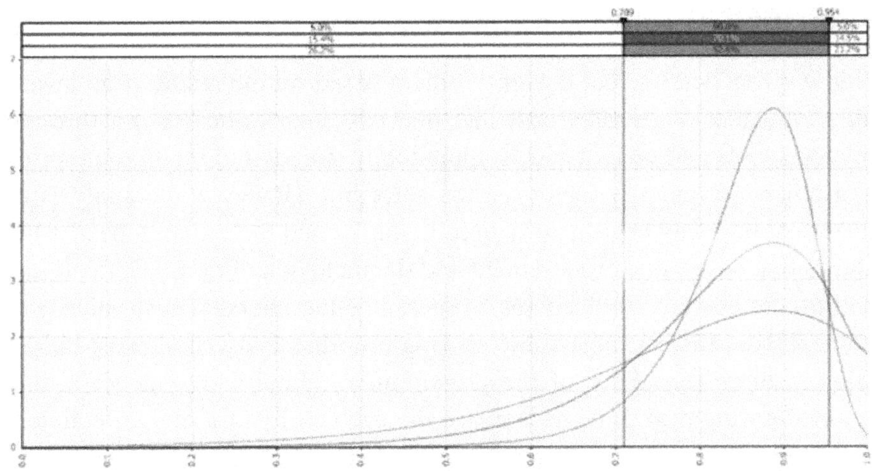

**Figure 6:** Probability distributions of non-cost scores for historical data (red) and Models 1 (blue) and 2 (green) used in the simulation.

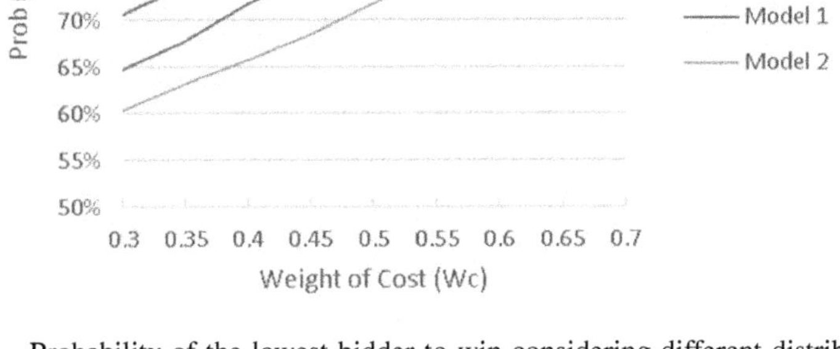


**Figure 7:** Probability of the lowest bidder to win considering different distributions of non-cost scores.

the range of scores used to evaluate technical proposals. Despite considering low values for the weight of cost (e.g., $Wc = 0.3$), the probability of the lowest bidder winning is higher than 50%, suggesting a high influence of cost scores skewness toward 1.

## Conclusions and recommendations

Highway agencies select design-builders based on the result of an award algorithm that aggregates cost and non-cost evaluation criteria through related scores and weights. This approach is intended to incorporate price and other relevant key factors in the selection. However, the analysis of historical highway projects procured in the United States using best-value shows that best-value procurement and the inclusion of non-cost factors in addition to cost do not necessarily guarantee the intended best-value outcome. Under historic best-value scoring criteria trends, the lowest bidder wins in more than 80% of the procurements. Intuitively, agencies may lower the weight of price to increase the probability of a higher technical score winning the project. However, this research shows that lowering the weight of cost to as low as 30% still results in a 70% probability of the lowest bidder winning the project. These results show that decreasing the weight of cost does not substantially increase the importance of non-cost factors in procurement outcomes. Therefore, agencies need to use a wider dispersion in scoring ranges if they want non-cost factors to influence best-value selection.

This chapter analyzes historical trends in scoring for highway projects in the United States to understand how current scoring may cause best-value procurement to be biased toward price. This analysis is important because the current bias of best-value procurement toward price generates a fallacy by which design-builders are incentivized to focus on price instead of seeking opportunities to provide value. The conclusions and recommendations related to scoring and weighting practices can be extrapolated to other countries and project scopes.

The main conclusion of this analysis is that adequate choices of weights and scores are needed to address the current fallacy in best-value procurement. The simulation approach presented in this study can be used to better understand the impact of these choices on procurement outcomes. Specifically, this study derives conclusions and recommendations aimed at helping agencies to do the following: (1) better understand the impact of scoring on best-value selection and (2) how to adjust weights based on expected bid dispersion.

The illustrative example presented in this chapter has demonstrated that evaluating non-cost criteria with a wider scale leads to increased probabilities of the highest-scoring proposal winning the project (e.g., "flipping" the bid from the lowest cost winning). Considering a wide range of non-cost scores in each procurement would enable agencies to make a more meaningful differentiation between the technical proposals. In other words, if all the proposers score equally in the technical evaluation (i.e., a narrow range of non-cost scores so that there is no "technical" distinction among proposers), the differentiator would not be the non-cost component but solely the cost.

From the analysis of current practices in highway best-value procurement in the United States, this study recommends broadening the range of scores used in the evaluation of the technical proposals. To determine how the non-cost scores range might be widened, agencies should consider a comprehensive analysis of their evaluation process. At a minimum, this analysis should include the following: (1) how the evaluation criteria are written, (2) the type of scoring system used and how it relates to the evaluation criteria, (3) the number of evaluators and the process followed to obtain the final non-cost score for each proposer, and (4) the evaluators' training to ensure the full scale of scores is used in the evaluation.

Agencies may be able to anticipate price dispersions based on project characteristics, as bid dispersion depends on the scope of the project and the risks involved. This information would be useful to predict whether the expected cost scores will be narrower or wider than the historical trends. This, in turn, can be used to inform appropriate weights for cost and non-cost factors. When low-bid dispersion is expected, the probability of having cost scores higher than non-cost scores increases. In these cases, agencies might adopt a lower weight of cost to better balance cost and non-cost factors. Similarly, if the bid dispersion is high, agencies may consider higher weights of cost.

In general, projects with a clear scope, low risks, and no substantial room for innovation can be expected to receive proposals with very close prices and, therefore, a narrow range of cost scores that will be close to 1. When procuring projects with an unclear scope, significant uncertain risks, and substantial room for innovation, agencies may expect to receive proposals with different prices and, therefore, high bid dispersion and wide ranges of cost scores.

In an analysis of US highway projects procured using best-value, Calahorra *et al.* (2021) proposed using a range of cost weights between 43% and 57% to balance cost and non-cost factors. Based on the results presented in this chapter and the recommendations in Calahorra *et al.* (2021), projects expected to have low-bid dispersion should be procured using low weights for cost scores (i.e., close to 45%), whereas projects expected to have high bid dispersion should be procured with higher weights for cost scores (i.e., close to 57%).

# About the Authors

**Maria Calahorra-Jimenez** is an Assistant Professor in the Department of Construction Management at California State University, Fresno. Maria holds a Ph.D. degree in Civil Engineering from the University of Colorado Boulder and a Ph.D. degree in Engineering Science from the Catholic University of Chile.

Maria's research focuses on improving the management of public infrastructure projects.

Before earning her doctorates, Maria worked in international engineering firms for 14 years. She served as a project engineer and project manager in the design and construction oversee of more than a hundred infrastructure projects — totaling nearly $2 billion — in Spain and Chile.

**Keith Molenaar** is the K. Stanton Lewis Professor of Construction Engineering and Management in the Department of Civil, Environmental and Architectural Engineering and the Dean for the College of Engineering and Applied Science at the University of Colorado Boulder. His research focuses on risk analysis and innovative project delivery methods. Dr. Molenaar holds a B.S. in Architectural Engineering and M.S. and Ph.D. degrees in Civil Engineering from the University of Colorado. Dr. Molenaar was elected into the National Academy of Construction (2017), the Pan-American Academy of Engineering (2012) and is a Fellow of the Design-Build Institute of America (2019).

**Cristina Torres-Machi** is an Assistant Professor in the Department of Civil, Environmental and Architectural Engineering at the University of Colorado Boulder. She has a dual Ph.D. degree from the Polytechnic University of Valencia in Spain and the Catholic University of Chile. Dr. Torres-Machi has more than 10 years of experience working on advancing the current management of infrastructure assets, with a focus on highway projects. Her research activities within the field revolve around deterioration modeling, optimization, and decision-making systems aimed to enhance the long-term condition and resilience of infrastructure systems.

# Chapter 9

# Technical and Organizational Challenges for Incentivizing Sustainable Public Procurement

Jeralee Anderson[1], Giovanni Migliaccio[2], Steve Muench[3] and Amlan Mukherjee[4]

[1]*Green Road International*
[2,3]*University of Washington*
[4]*Michigan Technological University*

## Introduction

The discussion in this chapter reviews the challenges of incentives for sustainable public procurement through the lenses of technical and organizational components. The technical component specifically focuses on the aspects of sustainable public procurement that relate directly to the development of novel engineering heuristics. Organizational challenges, on the other hand, address the creation of collaborative platforms and networks within which shared protocols and knowledge can support sustainable procurement practices. Ferraro, Etzion, and Gehman (2015) discuss that collaborative organization is necessary to tackle institutional change and cannot be brought forth by any one participant alone. Instead, change is created through collected, distributed action that translates new technical solutions into practice.

This chapter examines the technical and organizational challenges in integrating sustainable practices into public procurement. The project procurement process is an important pathway to incentivizing the adoption of sustainable practices in construction. A combination of instruments, policies, and practices including but not limited to Environmental Product Declaration (EPD), Best-Value Procurement (BVP), and incentive contracts can be used to shape the procurement process. Indeed, legislation such as the Buy Clean Act (2017) in California attempts to reduce long-term greenhouse gas emissions of public projects by requiring EPDs for eligible construction materials at the point of construction. Voluntary sustainability rating systems such as LEED, Green Globes, Greenroads, and Envision also encourage the use of a variety of sustainable practices that focus more broadly on the delivery of sustainable projects. The process of public procurement lies at the intersection of these different approaches, enacted through an informed policy that accounts for technical knowledge and organizational infrastructure.

This chapter reviews the current best practices and provides an introduction to the underlying principles for incorporating sustainability into public procurement for highway design and to inform sustainable construction practices (SCPs). The contents of this chapter summarize research on incorporating sustainability into public procurement as part of the National Cooperative for Highway Research Program (NCHRP) project on sustainable highway construction (Muench *et al.*, 2019) as well as the work from the Federal Highways Administration (FHWA) Sustainable Pavement Program (Mukherjee *et al.*, 2020).

## Incorporating sustainability into public procurement

Procurement is the process of purchasing the external services and materials necessary to deliver a project. As part of a recent workshop of agency and industry stakeholders from the construction industry, conducted in partial completion of the NCHRP 916: *Guidebook for Sustainable Highway Construction Practices*, the following organizational challenges to public procurement were identified:

- There is no standard process or documentation to incorporate sustainability into public procurement. There have been many sustainable/

green public procurement efforts, such as the United Nations' One Planet Network Sustainable Public Procurement (SPP) program (UN, 2020), the European Union's Green Public Procurement (GPP) criteria (EU, 2021), and the OECD's Going Green: Best Practices for Public Procurement (OECD, 2015). Still, there is nothing specific to roadway infrastructure.

- In construction project management, cost and schedule take routine priority over sustainability. There was little knowledge base on how to prioritize/emphasize sustainability through the usual channels of organizational support, pre-construction meetings, management plans, and providing feedback on sustainability efforts after construction. As a result, sustainability is not likely to be incorporated into project management goals and processes, and, even if it is, it tends to be lower than cost and schedule in priority.
- Isolating construction phase contributions from other highway sustainability phases is possible but not well documented. Comments tended to address planning and design elements, which are decisions about features and layout that are made upstream of procurement, and directly affect construction outcomes.
- A programmatic approach is better than one-off attempts because such an approach ensures sustainability is routinely considered on par with other priorities, and economies of scale can be achieved.

The remaining of this section provides guidance on including sustainability in current procurement methods. While some aspects in procurement already uses sustainability as a criterion (e.g., life-cycle assessment instruments discussed later), the emphasis in this section is on sustainable procurement of the entire project using best-value procurement, alternative technical concepts (ATCs), and sustainable procurement processes.

## *Sustainable procurement*

Sustainable procurement is a general term that describes a purchasing process that accounts for not only economic but also environmental and social considerations. In fact, this chapter is about sustainable procurement in the public sector. Globally, there is much guidance on sustainable procurement for goods and services, but specific construction guidelines can be lacking in detail. Some specific guidance, policy, and legislation examples involving construction are as follows:

- The European Union's *Revision of Green Public Procurement Criteria for Road Design, Construction and Maintenance* (Garbarino *et al.*, 2016) provides general guidance but stops short of recommending a specific language.
- The Construction Industry Research and Information Association's (CIRIA) *Guide to Sustainable Procurement in Construction (*Berry and McCarthy, 2011) provides guidance and case studies on sustainable procurement including how to apply BS 8903 (a British standard on sustainable procurement) to construction in particular.
- FHWA Order 4460.3A (2010) provides direction on programs that must be considered during purchasing (many of which can be considered to promote sustainability objectives). However, evidence of this Order's influence is minimal despite a comprehensive *FHWA Green Procurement Guide* (2010).
- The Buy Clean California Act (AB 262 signed into law in 2017) requires the establishment of maximum global warming potential (GWP) limits for select construction materials including structural steel, reinforcing steel, flat glass, and mineral wood board insulation. This necessitates the collection of Environmental Product Declarations (EPDs) on these products to document their GWP impacts. Caltrans has also conducted pilot projects that include these items and concrete, asphalt, and aggregate. This effort focuses on procuring certain construction materials, but it does not address the procurement of the project itself.

## *Best-value procurement*

Best-value procurement (BVP) is a procurement process that considers both qualitative and quantitative criteria for selecting project contractors and vendors. In a BVP process, it is possible to include sustainability criteria in those selection criteria. While NCHRP Report 561 (Scott *et al.*, 2006) recommends an extensive list of criteria to be included as BVP evaluation, the following could also be included specifically as sustainability criteria:

- Include an environmental considerations evaluation criterion.
- Ask for proposed practices that address the environment and go beyond meeting regulatory environmental requirements.
- Include a life-cycle cost evaluation criterion. Focus on portions of the project where life-cycle cost procedures are well defined based on existing literature. For instance, there are published standard practices for life-cycle costing of bridges, tunnels, and pavements.
- Set a project goal of achieving a sustainable rating system score, rating, or achievement level. A BVP evaluation criterion could help the proposers to contribute to the goal's achievement. It is appropriate to ask proposers to identify specific rating system credits, their value, and the specific construction means and methods needed to achieve the credits. These commitments can result in the proposer investing in sustainability achievement *so long as the commitment and achievements are measured and managed.*
- Use lane rental and traffic control evaluation criteria to incentivize shorter construction schedules to reduce user costs.
- Reward contractors who have demonstrated experience with sustainability efforts or have credentialed personnel on staff and assigned to the project.
- Require a means of tracking sustainability commitments through the project. Such a sustainability management plan can be used to track progress, achievement, and changes to sustainability goals throughout the project life like a schedule tracks timing/effort/coordination, a budget tracks cost, or a safety management plan tracks safety.

Guidance for each of the above approaches can be found in the *NCHRP Report 916: Guidebook for Sustainable Construction Practices.*

To date, where sustainability has been included in BVP it is minimally impactful because either

1. it is included as a non-specific portion of a technical solution evaluation criterion,

2. it is essentially a measure of how well the project meets the environmental regulatory requirements, or
3. it is assigned a relatively minor factor/weight (often 5% or less of the scored criteria).

When using sustainability as a BVP technical solution evaluation criterion, it may be prudent to first score it as a small fraction of the total qualification and/or technical score to allow owners and proposers to adjust to the new requirements while it has a relatively small influence on the contract award. An initial target of 5–10% of qualifications and/or technical score is recommended by the authors. As familiarity grows, this fraction can grow as appropriate (for instance, on a project with high/multiple sustainability priorities).

### *Alternative Technical Concepts (ATCs)*

The FHWA defines Alternative Technical Concepts (ATCs) as "suggested changes submitted by proposing teams to the contracting agency's supplied basic configurations, project scope, design or construction criteria. These proposed changes provide a solution that is equal or better to the requirements in the Request for Proposal document. If the ATC concept is acceptable to the contracting agency, the concept may be incorporated as part of the proposing teams technical and price submittal. ATCs provide flexibility to the proposers in order to enhance innovation and achieve efficiency." ATCs are meant to improve a project, typically by reducing cost, reducing construction time and/or disruption, or improving environmental outcomes. It is possible to include sustainability as an ATC evaluation criterion but the owners must precisely define the criterion (as opposed to a general undefined request for "sustainability") to receive meaningful consideration.

## Life-Cycle Assessment-based approaches

As the science of Life-Cycle Assessment (LCA) has advanced so has the public's understanding of its usefulness in supporting decision-making. This has led to pressure on regulatory bodies in considering the use of

LCA and Environmental Product Declarations (EPD) in informing public procurement. Legislatively, the Buy Clean California Act, 2017, (Assembly Bill 262), is requiring successful bids on public construction projects to produce EPD for a list of eligible materials at the point of installation. Similar bills have been introduced in Washington and Minnesota. In this context, it is imperative for state departments of transportations (DOTs) to proactively identify the best strategies for integrating EPDs into the construction project procurement process. Broadly, this section addresses the three following questions:

- What are the current approaches to ensuring there are reliable consensus-based LCA heuristics to support sustainable decision-making?
- What is the current state of legislative and administrative infrastructure (current and pending legislation) that supports the use of LCA-based instruments to support sustainable decision-making?
- Do we have the statistical methods to use LCA outcomes when using them to compare the sustainability of alternative products and processes?

This section starts with a discussion of the ISO standards and relevant LCA-based instruments that are part of the discussion in sustainable decision-making. The role of the Sustainable Pavements Program as a collaborative organizational platform in furthering the development of consensus-based LCA methods is discussed with pointers to its various research and development products. Next, a discussion of the Buy Clean Act (2017) in California is included to highlight how a legislative measure brings together the need for technical and organizational infrastructure in supporting LCA-based decision-making. Finally, the chapter concludes with an assessment of the ability for LCA-based methods to address engineering questions around trade-off and scale that are crucial for the support of sustainable decision-making.

## *Relevant standards and instruments*

The International Organization for Standardization (ISO) has established standard processes and procedures for developing product labels that use methods in Life-Cycle Assessment (LCA) to declare the environmental impacts of construction materials [(ISO 14040:2006), (ISO 14044:2006), (ISO 14025:2006)]. An Environmental Product Declaration (EPD) is a Type III Environmental Declaration that communicates the potential environmental impacts of a product or service using methods in Life-Cycle Assessment (LCA) (ISO 14025:2006). An EPD is a very useful instrument for communicating the outcomes of an LCA. The specific strengths lie in the rigorous process required by standards such as (ISO 14025:2006) and (EN 15804:2012). The clear definition of the goal and scope of the underlying LCA in the PCR ensures consistent use of system boundaries and functional or declared units.

LCA outcomes in EPDs are typically communicated using mid-point indicators including potentials such as global warming (GWP) (kg of $CO_2$ Eq.), acidification (kg of $SO_2$ Eq.), eutrophication (kg of N Eq.), smog formation (kg of $O_3$ Eq.), ozone depletion (kg of CFC-11 Eq.), and particulate matter (PM2.5). In addition, resource depletion and the life cycle energy consumption metrics are included as well. The calculation of LCA heuristics is fundamentally a technical problem that has well-developed solutions. However, integrating LCA outcomes into the decision-making processes particularly to support public procurement opens up a set of problems that cannot be solved without accounting for their organizational contexts.

In its essence, LCA is fundamentally an attempt at assessing the environmental impacts of a product or process using a model of the underlying supply chains using estimates used from multiple products and processes. As with any model, LCA results are often sensitive to multiple assumptions. The communication of LCA outcomes using EPDs that are compliant with industry- and product-specific PCRs was a crucial step toward the standardization of the goal and scope of the LCA and the assumptions that are part of the definition of system boundaries. In addition, given the interconnectedness and interdependence of supply chains of construction materials, this consensus must be extended to include a harmonized consensus on assumptions across different industries. The Sustainable Pavements Program discussed in the next section is an example of a collaborative organization that served as the platform for the development of a consensus-based approach necessary to support LCA as a tool for supporting pavement sustainability.

## *The sustainable pavements program*

The inception of the Federal Highways Administration (FHWA) Sustainable Pavements Technical Working Group (SPTWG) in 2011 as part of the FHWA Sustainable Pavement Program (SPP) provided the organizational platform for the pavements industry to develop consensus around LCA of pavement design and construction and the associated products and processes. The work of the SPTWG led to the Pavement LCA framework and the FHWA pavement sustainability reference manual (Van Dam *et al.*, 2015) — a useful resource for state and federal agencies to develop a better understanding of LCA and its relationship to building sustainable pavements. Some agencies are starting to follow a structured approach to developing their own LCA strategies, for example, Caltrans is considering the use of EPD in construction procurement and Illinois Tollway has developed an LCA-based tool for project evaluation.

An outcome of the discussions at the SPTWG meetings has been the development of the pyramid that provides perspective on the organization of the LCA framework and the road ahead. Figure 1 illustrates the pyramid. At the very base of the pyramid is the LCA framework development that has been completed. The next level considers the requisite data definition that will support the implementation and easy use of the LCA framework. The importance of the data definition layer in the pyramid is

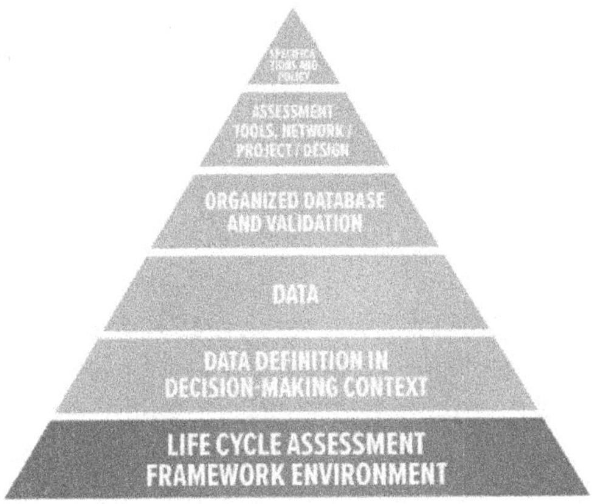


**Figure 1:** The pyramid.

to create a common vocabulary for collecting, exchanging, and using pavement life cycle data. The primary idea conveyed by the pyramid is that while the development of an LCA tool — a strictly technical problem — can be achieved relatively easily, the validity of the tool and therefore its usefulness is a function of the consensus models that provide the foundation of the pyramid.

The pavement construction materials industry has embraced LCA and both the concrete and asphalt industries have developed ISO 14025 and EN 15804 compliant Environmental Product Declarations (EPD) using ISO 21930 compliant EPD programs. Academic inquiry has supported these developments through support of industry product category rule (PCR) development and furthering models of the use stage of Pavement LCA, with emphasis on topics such as pavement vehicle interaction (PVI) and heat island effect.

The FHWA supported a workshop in September of 2016 to address infrastructure life cycle inventory data needs with the goal of supporting sustainable decision-making using EPDs. The workshop was well attended with representation from a diverse set of stakeholders in the pavement industry. The recommendation of the workshop was to develop standardized data structures that can be used to collect, report, and compute datasets for Pavement LCA. A specific recommendation was to identify and harmonize common background datasets and protocols that are used across different processes and product systems to ensure their consistent and reliable use.

A culmination of the SPP's collaborative efforts was the development of a Pavement LCA Tool, in keeping with the FHWA's Pavement LCA Framework. The objective of the tool is to help agencies begin assessing the environmental footprint of their decisions using LCA principles. A critical component of this tool is the use of reliable, transparent, and comprehensive life cycle inventory (LCI) data. LCI data includes foreground data that is directly observed from processes being studied and collected as part of the LCA and background data (or upstream inventories) that are often used from industry averages. This tool is easy to use and can be used by agencies to use LCA information to inform design and construction alternatives.

## *Use of LCA and EPDs in policy*

A detailed discussion of the technical and organizational challenges of using LCAs and EPDs in legislation and policy can be found in Mukherjee *et al.* (2020). This section highlights primary takeaways.

In the public procurement space, EPDs have been identified as the primary LCA-based instrument, as has been indicated by their use in legislation in California, the Buy Clean Act (2017). The Bill broadly requires awarding authorities (such as Caltrans) to do the following (Buy Clean California Act 2017):

- Require submission of EPDs by successful bidders before the installation of any "eligible material". A list of eligible materials has been provided, and a project with any eligible material in it is considered an "eligible project".
- Accept EPDs that are compliant with "International Organization for Standardization (ISO) standard 14025, or similarly robust life cycle assessment methods that have uniform standards in data collection consistent with ISO standard 14025, industry acceptance, and integrity, for each eligible material."
- Set a "maximum acceptable global warming potential as a number that states the maximum acceptable facility-specific global warming potential" for each eligible material by January 1, 2019. The estimates will be based on industry averages by consulting, "nationally or internationally recognized databases of environmental product declarations".
- Review the maximum acceptable global warming potential by January 1, 2022, and every three years thereafter, and revise them downward to reflect industry improvements to ensure continuous improvement.

While the legislative intent of the legislation is clear, its impact on current business processes within agencies that are directly involved in the procurement of eligible materials on eligible infrastructure projects such as Caltrans is yet to be completely documented. At the time of writing this article, Caltrans was in the process of establishing processes that account for the technical and organizational challenges in implementing EPD in the decision-making process. Copycat bills similar to the one supporting this kind of legislation have been introduced in a few other states including Washington and Minnesota. While neither state has, at the time of this writing, adopted similar legislation, both states have completed initial studies (Simonen, 2019; Mukherjee and Miller, 2019).

In summary, some of the challenges that agencies will face as identified from Caltrans' experience in implementing AB 262 are as follows:

1. Ensure consistent reporting of life cycle inventories that are used to conduct LCAs that support material EPDs being produced at the point of installation.
2. Ensure harmonization across PCRs of eligible construction materials. The lack of harmonization of PCRs can result in material flows being excluded from accounting or, in some cases, double counted, thus reducing the reliability of the EPDs produced.
3. Develop infrastructure to appropriately use EPDs including the development of the maximum acceptable limits for GWP and the contexts in which they are applied.

As discussed earlier, these challenges need to be solved collaboratively with the recognition that they must include interdependent technical and organizational components. Agencies will need to develop their individual roadmaps to implementing the use of LCA/EPD in the highway construction procurement process by identifying ways to best address fundamental technical and organizational challenges. Workforce development in the construction contracting industry with tools and resources to produce EPDs/LCA that are compliant with agency needs will also need to be a priority.

## *Use of LCA and EPD in decision-making*

At the heart of using LCA in supporting sustainability is the need to quantitatively differentiate between alternative designs and methods. At a very high level, LCA can be used to inform pavement design selection decisions or help select the best maintenance and rehabilitation strategy for a given pavement section. At present, LCAs are primarily being used for informational purposes to create transparency rather than to inform decisions. However, in future, the intention remains to use LCA as the quantitative approach to discerning between the life cycle sustainability of alternatives.

This raises the challenges around carefully considering the decision contexts within which the LCA/EPD information are being used. An important challenge is the role of uncertainty in LCA and the limited understanding of the margins of error that are acceptable when developing estimates for specific design alternatives. Another way to address this same problem is to develop ways to assess the level of significance in a unit difference in GWP between the two options. For instance, if the acceptable error margins for benchmarking GWP for two similar options are +/–X% and +/–Y%, respectively, then for some range that is a function of X% and Y%, the options should be considered effectively equivalent. While some preliminary approached have been developed in this topic (Bhat and Mukherjee, 2020), there remains significant gaps in developing methods to support best practices. This also establishes the necessity of appropriate statistical aggregation and analysis in the development averages and maximum acceptable limits. A specific context in which this directly applies is when comparing a facility-specific EPD to a national industry average. Given the climatic diversity in North America as well as differences in regional materials and preferences, such a comparison can only be conducted fairly after developing confidence intervals, regional factors, and reasonable margins of error.

Another aspect of using LCA in decision-making is to consider the design contexts of the materials being used and the implications that may have for decision-making. As most LCA instruments in the context of procurement are limited to a cradle-to-gate scope, it is crucial to consider the impact of trade-offs between environmental impacts and long-term performance of pavements. For example, when using recycled materials, an environmental benefit may offset long-term performance, while the use of additives and polymers can improve long-term performance while potentially increasing the life cycle impacts. Trade-off mechanisms will have to be developed to ensure suitable handling of these contexts (Mukherjee and Bhat, 2019; Qazi *et al.*, 2014).

## Contractual instruments and specifications

Contracting is the process of establishing a legally enforceable agreement expressing the expectations, responsibilities, and protections of each party for services and materials. Regarding sustainability, the contract is the primary vehicle through which sustainability goals and requirements are communicated, associated risk is allocated, and payment is made or penalties are assessed based on achievement. In the US, sustainability is incorporated into roadway construction when contracts allow, mandate, and/or incentivize it.

### *Allowing sustainability*

The most passive contractual approach to sustainability is to allow sustainable practices but not require them. It is the least effective approach because there is no guarantee that any sustainable practice will occur. Typically, a contractor will only implement an allowed, but not required, sustainable practice if it is, independent of the contract, a business opportunity, meaning it can increase revenue or market share, reduce expenses, increase employee productivity, or reduce risk. An example of allowing a sustainable practice is the use of reclaimed asphalt pavement (RAP) in asphalt pavements. Most state departments of transportation (DOTs) specify an allowable maximum amount, but do not mandate its inclusion. Thus, contractors use RAP when it is advantageous to their business to do so, but they are not obligated to use it.

### *Requiring sustainability*

The most aggressive contractual approach to sustainability is to require enumerated sustainable practices within the contract. One way to do this is to specifically call out each sustainable practice in the contract and specifications as would be done for any other construction practice. Alternatively, a performance-based metric can be identified in the contract (for instance, a requirement to achieve a certain certification level in an external rating system), allowing the contractor to meet it by choosing from a variety of sustainable practices. This provides contractors with the flexibility to meet sustainability requirement in their most cost-effective and efficient manner.

### *Incentivizing sustainability*

A positive contractual approach to sustainability is to provide incentives for the inclusion of sustainable practices. In providing incentives, owners should be aware that contractors usually need to spend money to achieve an incentive, so sustainability incentives should not be viewed as entirely supplemental income/profit to the contractor. Incentivizing sustainability is between allowing and requiring it in that the contractor still has some choice about pursuing specific items, but can be financially encouraged to pursue those items. Still, there is no guarantee that any incentivized sustainability action will be accomplished, since accomplishment is at the contractor's discretion. An example of incentivizing a sustainable practice is an offered payment from an owner to a contractor if the contractor's project team has one or more personnel credentialed in the sustainability rating system being used on the project.

### *Pricing sustainability*

Sustainable practices can be priced separately as an option or alternate bid. In this manner, the owner can, conceivably, isolate the prices for specific sustainable practices as an alternate bid which is a defined portion of the work that is priced separately, which provides the owner an option to include that scope. In this way, a baseline bid can be compared with a sustainable alternate bid. Even if done in this manner, the cost of sustainability is still difficult to determine because bids indicate price, not cost, and many sustainable practices have little history making accurate pricing difficult for the contractor. For instance, a contractor may price a sustainability management plan at a particular level, which may be because the contractor is unsure of the work an owner will require for such a plan and therefore prices in a substantial amount of risk or the contractor has no history with sustainability management plans, therefore prices in development work that amounts to a one-time cost that can be spread over multiple jobs.

## *Sustainability risk*

Contracts determine risk allocation between participating parties. Risks involved with sustainable practices may be less understood and quantifiable than other, more traditional risks. For instance, there is a risk that although a contract item may function as intended, it could still not achieve the sustainable outcome required. Or else, that same item may be removed from the contract by change order, but it is needed to meet a rating system certification or other sustainability metrics.

Project delivery methods that make communication amongst parties easier, such as design–build or integrated project delivery, are generally better suited to sustainable procurement and contracting because sustainability requirements are relatively new, requiring more communication to agree on processes and details. Korkmaz *et al.* (2010) found that project delivery systems that emphasize sharing of project risks in a non-adversarial manner, such as systems based on the design–build family, show an improved performance on project sustainability metrics.

## *Sustainability management plans*

A sustainability management plan is a documented plan that describes how sustainability will be incorporated, measured, managed, and reported within an organization or project. Typically, such a plan would describe the project's sustainability objectives, commitments, measurements, and management process as they relate specifically to construction. In this way, high-level project sustainability goals can be translated into specific construction goals and practices, and progress toward them can be measured and managed. If used, a sustainability management plan should be required by specification, part of contract documentation, and a construction meeting topic to ensure it is given proper importance. Currently, there are no sustainability management plan standards and few examples to follow.

## Green rating systems

Sustainability (or "green") rating systems are collections of metrics for environmental, social, and economic performance that serve as an indicator for overall sustainability on a project. When successfully executed, the performance measures represent a collective sustainability benefit for the project overall. Sustainability rating systems can be used at any time in the development process and are especially useful in purpose and needs analysis during early planning and later during construction planning and procurement. Rating systems assist with sustainability goal setting, development of a collaborative and interdisciplinary team, and establishing tracking for environmental, social, and economic commitments in construction. Sustainability scores also influence document control practices, public and stakeholder engagement, and more.

Most rating systems are voluntary and scores or certifications are optional pursuits that provide validation from an outside authority perceived as subject matter experts. Currently, sustainability ratings (scores) are generated from two categories of tools: self-assessment and third-party tools. Rating systems also vary in applicability to project phases (planning, design, construction, and operations), how activities are weighted, level of scientific study and calibration, level of difficulty or effort required, amount of documentation required, transparency of weighting, independence of creators, and scope or discipline that a particular tool supports. There are many types of rating systems, and it is up to the user to understand how to apply each tool and to be able to identify the appropriate context for its use.

Rating systems can incentivize better and more innovative behaviors that lead to positive sustainability outcomes and community benefits because the tools provide recognition and validation of these activities that pure compliance with regulation does not. Most rating system tools are continuously evolving, consensus-based standards that provide a clear target or goal as a motivator for incorporating sustainability practices.

The sustainability rating system experience is a differentiator in the highway construction industry where currently this experience is scarce or absent in many parts of the world. BVP as well as other forms of alternative delivery promote integrated teams that are more collaborative and as a result have scores that are significantly higher than traditional procurement methods (Anderson, 2013). Contractors, manufacturers, and suppliers with sustainability experience, as shown through demonstrated support of or participation in a rating program or by educating or hiring accredited staff, tend to win more projects where sustainability is a requirement or a request in the procurement process.

Rating systems, and the measures in them in whole or in part, can be incorporated into construction contracts in different ways. For example, the project owner can determine a particular score is desired and it can be required. Or else, a project team can propose to deliver a certain score or rating. Once the team determines the appropriate rating tool and target, roles and procedures can be outlined and clarified early in project procurement to understand the different responsibilities and resources required to achieve the desired overall rating or score.

## Policy implications

Procurement policy is an operational policy that impacts different departments or services and may not be a legislative initiative or public policy. When the government buys products and services from others, it defines codes for each type of product or service for accounting and purchasing of certain approved products and services. As of October 2020, the US General Services Administration Federal Acquisition Services (FAR 2.101) defines "environmentally preferable" in its procurement policy to identify (GSA, 2020):

> *products or services that have a lesser or reduced effect on human health and the environment when compared with competing products or services that serve the same purpose. This comparison may consider raw materials acquisition, production, manufacturing, packaging,*

> *distribution, reuse, operation, maintenance, or disposal of the product or service. For the Product Services Code Manual, this includes categories such as EPEAT registered, water efficient, Significant New Alternatives Program (SNAP)/non-ozone depleting, recovered content, and environmentally preferable products/services (EPA designated).*
>
> *For the Product Services Code Manual, environmentally preferable also includes building-related or facilities related services which are associated with "green" industry standards and certifications, such as the US Green Building Council's (USGBC) Leadership in Energy and Environmental Design (LEED) certification. Environmentally preferable excludes products or services that are predominately bio-based or energy-efficient as these two categories are defined in the Products and Services Code Manual.*

These procurement requirements are shown by the agency on any requests for proposals and may include specific rating systems or minimum product requirements.

At present, the primary instruments that are being *required* in procurement are LCA-based instruments such as EPD (these are also referred to as eco-labels, ISO Type III) or through *voluntary* participation using rating systems.

## LCA-based instruments in legislation

Laws such as the Buy Clean Act impact the market of materials. However, communication to all stakeholders regarding how the information provided in the EPDs will be used to inform business processes will improve industry and agency outcomes. At present, the EPDs are being collected only for informational purposes though there is a possibility that in the future, they will support decision-making. A plan for appropriate use of EPDs, along with clear communication of it can go a long way in ensuring long-term adoption of EPDs in practice. See detailed discussion of Buy Clean Act in Section 3.

Moving forward, a federal implementation of legislation like the Buy Clean Act through governmental contracting organizations such as the General Services Administration will significantly support sustainable public procurement.

## *Rating systems in legislation*

Rating system tools have broad potential effects for sustainable procurement in legislation and also influence the construction industry. Rating systems can be part of development permit criteria as an explicit incentive for a public benefit, a local resolution or ordinance, land use and master plan commitments, or state or federal law. In particular, rating systems can be a particularly effective tool for tying sustainability performance to legislative efforts like budgets, when used in conjunction with or as a replacement for funding prioritization criteria of capital projects.

Blanket legislation or public policy of any kind serves to make a voluntary rating system mandatory and raises the performance expectations locally. A mandate tends to increase the stakes for successful delivery of a sustainable construction project by removing the incentive of voluntary engagement entirely, which may result in only minimum performance being met unless additional incentives (such as performance bonuses) are also included in procurement. A policy mandate or commitment becomes a means to raise the norm to a new level. It follows that a mandate for use of a rating system or similar tool tends to drive innovation by influencing the standard setters themselves by requiring continual updates to the referenced standard and increasing familiarity with the rating system internal to an agency for executing it and increasing public support and awareness, which increases public pressure and expectations.

Moving forward, significant work remains to be done in integrating procurement policy with the underlying science of LCA and the experience of rating systems, to create the impact necessary for achieving long-term sustainability goals.

# About the Authors

**Jeralee Anderson** is the CEO and Co-Founder of Greenroads International, a globally recognized non-profit organization leading the green and equitable transformation of the transportation industry and infrastructure through education and the independent third-party sustainability certification of transportation projects. Dr. Anderson is a licensed professional engineer in Washington State and California and was recognized in 2013 by the Obama White House as a Transportation Champion of Change for the 21st century. In 2018, Jeralee was elected to the Redmond City Council, serves as the Council Vice President, and in 2019 was appointed to serve on the Washington State Public Works Board by Governor Jay Inslee. She also represents the City of Redmond as an alternate on the King County Regional Transit Committee and the Cascade Water Alliance.

**Giovanni C. Migliaccio** holds a Howard S. Wright Endowed Professorship, he is the current Chair of the Department of Construction Management, and he has served as the Executive Director of the Center for Education and Research in Construction (CERC). He holds an M.S. and a Ph.D. in Civil Engineering from the University of Texas at Austin and a master-level degree from Politecnico di Bari in Italy. Prior to moving to the US, he worked in Italy in the construction management of telecommunication projects under Nortel Networks, Nokia Networks, and IPSE 2000. At UW, Giovanni is active in all three pillars of academic life, including teaching, research, and service to the university and the industry.

**Steve Muench** is the Tom and Marilyn Draeger | The Beavers Charitable Trust Professor in the University of Washington's Department of Civil and Environmental Engineering. His research interests include sustainability, roads, construction, and learning. Steve has helped spin off two companies based on his research at UW: Greenroads International and HeadLight. He is a licensed professional engineer, and before his academic career Steve spent two years as a transportation design engineer and seven years as a US Navy nuclear submarine officer.

**Amlan Mukherjee** is a Professor in the Civil, Environmental & Geospatial Engineering Department at Michigan Technological University. His research interests have primarily been in studying civil infrastructure systems through the lens of risk and life cycle analysis, to improve project outcomes such as cost, performance, and environmental impacts. He has extensive background and experience in life-cycle assessment for the construction materials industries and has facilitated the development and implementation of the Environmental Product Development Program for North American asphalt mixtures for the National Asphalt Pavement Association. He has also served on the Federal Highways Administration (FHWA) Sustainable Pavements Technical Working Group. In related work, he has co-authored the NCHRP guidelines for sustainable highway construction practices. Dr. Mukherjee also serves on the Board of Directors for the Green Buildings Initiative and the Greenroads Foundation. He received his Ph.D. from the University of Washington, and is a licensed Professional Engineer in the State of Michigan.

# Part C
# Contracting

# Chapter 10

# Traditional Contracting Approaches for Design, Construction, and Materials

Giovanni C. Migliaccio[1], Pramen P. Shrestha[2] and G. Edward Gibson Jr.[3]

[1]*University of Washington*
[2]*University of Nevada Las Vegas*
[3]*Arizona State University*

## Background

Once a new project participant is selected, the contracting period begins, although drafting the contract documents is started much earlier. Signing the contract establishes a contractual relationship for services and materials through a written agreement that expresses the expectations, responsibilities, and protections of each party. Contracting for construction services is substantially different from contracting for design services or for purchasing materials.

Traditionally, the process of establishing a contractual relationship for design services is often based on reciprocal trust between parties and creates a fiduciary or "arm in arm" relationship. The purchasing party trusts the technical expertise of the designer, whose selection is usually based on qualifications. Moreover, design contractual documents often create an agency relationship by authorizing the designer to act as an agent on behalf of the client for various decisions, such as permitting or design interpretation. Agency relationships are implicitly trust-based. Contracting for construction management and preconstruction services retains many of the elements of design contracting, including the agency relationship.

Conversely, the process of establishing a contractual relationship for construction services is based on an arm's length transaction, which legally means the two parties are working in their self-interest during the contracting independently, while being free of reciprocal pressure or duress. As a result, construction contractual documents are expected to be grounded in design developed at the date of the contract and crafted to allocate risks among contractual parties, all while being thorough. Design–build contracting retains many of the elements of construction contracting, but it often results in producing contractual documents that more flexibly adapt to the project objectives. Contracting for materials is usually a purchasing act, as the entity providing these materials would not often be involved in their installation. However, in those cases where installations are performed by the material vendor, a contract for specialty contracting is usually adopted to incorporate both the materials and their installation.

## Construction contract documents

Construction contracts consist of the contracting parties' agreements related to many aspects of a project, including scope of work, consideration, and risk allocation. The contracts are drafted to reflect the goals and expectations of the parties. Contracts usually start with the owner's standard terms and conditions, followed by general terms and conditions, special conditions, terms of payment, drawings, specifications, and any other technical requirements for the specified work.

A contractual relationship for construction services and materials relies on a set of written documents. The main document is a contractual agreement. Through a legal concept called incorporation by reference, the construction contractual agreement is also used to incorporate the other contractual documents which include the following:

The *general conditions* document, which describes operating procedures that the owner usually uses on all projects of similar type.

The *special conditions* document, which describes operating procedures that are unique requirements for the project.

The *construction drawings*, which provide a graphical description of the project, including its geometrical information and often its materials.

The *construction specifications*, which provide a textual description of the project, including any additional information on its materials, quality acceptance processes or other performance expectations. Construction specifications are usually organized according to a standardized classification system. For instance, the Construction Specification Institute (CSI) Master Format is a commonly used classification system for organizing information about construction requirements, products, and activities.

To incorporate other documents, the contractual agreement must refer to them, provide them as attachments, and clearly state they are included as part of the contract. In some cases, publicly available regulatory standards and codes are simply cited. This approach allows for the organization of the rights and obligations of each signatory party into a set of documents that describes what needs to be built, how it needs to be done, and when it needs to be ready. The "what" is usually defined in the scope of work, which is summarized in the contractual agreement and detailed in the construction drawings; the "how" are the methods and quality requirements, which can be identified by reviewing the project specifications and conditions; the "when" will be stated in the contractual agreement. At a minimum, a construction contract includes information listed below:

- Who: the signatory contractual parties and any relationship to other parties.
- What: the work to be performed.
- Where: the location of the project.
- When: the time allowed to perform, including any financial disincentive for late completion, also known as liquidated damages.
- How much: the pricing method, including contract price, if known, or the process to evaluate costs to be reimbursed and fees to be awarded.
- How to get paid: the payment method, including processes to obtain progress and final payments.
- Other "Hows" would typically include the following: processes to address issues that may arise during contract administration,

including changes to the contract, disputes, unforeseen circumstances, design flaws, and contract terminations.

## Statutory framework regarding construction contracts under various delivery methods

When using alternative PDMs (e.g., DB, CMAR, and PPP), a mere revision to a traditional contract document tailored to the chosen PDM would not be adequate. The contract would lack key provisions required for the successful completion of projects using the above-mentioned alternative delivery methods. Many architectural, engineering, and construction trade groups (e.g., AIA, EJCDC, and Associated General Contractors as well as governmental agencies) have drafted specific contract documents tailored to the unique aspects of various PDMs. These organizations drafted PDM-specific contracts to address the unique characteristics of each method as well as the organization's specific policies, procedures, and applicable legal requirements and limitations. Since DB and PPP integrate design with construction, it is particularly important to note the various statutes that govern the practice of engineering and construction. Such statutes could complicate the use of alternative delivery methods. The major issues that raise complications in these delivery methods are related to construction firm's leadership role in a lump-sum DB team, professional liability, insurance coverage, or the lump-sum DB firm's inability to obtain its own errors and omissions coverage, among many others (ACINA, 2012).

Private entities can use any form of construction contracts. However, public entities have been limited to using a construction contract prepared for DBB projects. Two reasons for this limitation are: (1) the numerous laws and statutes that limited the procurement of construction services to the lowest responsible and responsive bidder; (2) the Brooks Act enacted in 1972 (Public Law 92-582), which limited the procurement of design services on federally funded projects to a qualifications-based selection method.

Construction contracts incorporate procurement rules of their funding organizations. These are either federal, state, municipality, or private entities. For instance, the use of the DB for federal projects under highways, airports, and other infrastructures was approved only after the Federal Acquisition Reform Act was enacted in 1996. In addition to the construction contract, however, the procurement of professional services and materials under these DB methods was also formalized after the introduction of this act.

Regardless of the PDM being utilized, for projects that are supported by federal funding, owners still need to follow the applicable federal procurement and contracting standards outlined in the Code of Federal Regulation (CFR). This includes 'CFR 18.36: Procurement' for agreements executed prior to December 26, 2014 and '2 CFR Part 200: Uniform Administrative Requirements, Cost Principles, and Audit Requirements for Federal Awards' for new awards and additional funding to existing awards executed on or after December 26, 2014.

## Standard forms of agreement

Some documents — special conditions, construction drawings, and construction specifications — are highly contextual to the project, whereas the contractual agreement and general conditions lay a legal framework that can be enforced in case of disputes. As a result, these documents are often standardized and many versions that exist are produced by either industry associations, public agencies, or private organizations with a significant capital project program. These sets include documents, sometimes improperly named standard contracts, that can be adapted to a specific project by filling in the blanks in the template. Some sets of standard forms of agreement below are most frequently used by owners who do not use a proprietary set of documents:

- AIA Contracts developed by the American Institute of Architects (AIA).
- ConsensusDocs developed by 40 leading associations in the construction industry.
- NAHB Contracts developed by the National Association of Home Builders (NAHB) for residential projects.
- Engineers Joint Contract Documents Committee (EJCDC) Contract Documents developed by a coalition of three major organizations of professional engineers, including The American Council of Engineering Companies, The National Society of Professional

Engineers, and The American Society of Civil Engineers – Construction Institute.

- Standard forms of agreement also exist internationally, such as Joint Contracts Tribunal (JCT) and the New Engineering Contract (NEC) in the UK, the Fédération Internationale des Ingénieurs-Conseils (FIDIC) in Europe and Asia, and so on.

For example, the American Institute of Architecture (AIA) has developed standard forms of documents for both the contractual agreement and the general conditions. While the General Conditions of the Contract for Construction (A201-2017) document is applicable to all projects independently from the pricing approach, specific versions of the contractual agreements have been developed for specific pricing approaches, such as the Standard Form of Agreement between Owner and Contractor (A101-2017). This form is used when the goal is to establish a contractual agreement between the owner and contractor under a stipulated sum approach, also known as a lump-sum or fixed price contracting approach. This form is also designed to be used with the A201-2017 general condition document. The A201-2017 document consists of the contract's general conditions for the construction of a large project. It includes the rights, responsibilities, and relationships between owner, contractor, and architect. While the architect is not a contractual party in the A101-2017 construction contract between the owner and contractor, the architect usually assists the owner in preparing the contract documents and performs some duties and responsibilities during the construction phase. The AIA contract document series also includes standard agreements for other forms of delivery, such as design–build and integrated project delivery or pricing. Such pricing may include the cost of the work plus a fee with or without a guaranteed maximum price.

Another example is provided by the standard documents published by the Engineers Join Contract Documents Committee (EJCDC), which are tailored for infrastructure projects. EJCDC publishes four series of documents, and they are construction series, engineering series, design–build, and procurement. The construction series consists of 25 documents, and it assumes that the design engineer is involved during construction. The second document series consists of standard conditions for the contract between the owner engineer and the engineer subconsultant. The third document series is developed specially for design–build projects and includes 17 design–build documents, including agreements (lump-sum and cost-plus), general conditions of contracts, performance bonds, owner–design–builder contracts, design–builder–engineer contracts, design–builder–subcontracts, and public–private partnership contracts. The final document series is specifically used for the procurement of infrastructure projects.

## Types of construction contracts

Information to be contained in a contractual agreement is highly dependent upon the project and upon the parties that are signatory to the agreement. The distinction between prime contracts and subcontracts is important when parties are considered and is described hereafter. Still, contractual terms can vary largely among projects with the exception of the pricing and payment methods. In fact, pricing methods for construction contracts mostly fall within three types: (a) cost plus a fee (with or without a guaranteed maximum price); (b) unit prices; (c) lump sum; these methods are also described. The pricing method often affects the payment method. Similar to the project delivery and procurement methods, the owner also selects the project pricing and payment methods for prime contracts.

### *Prime contracts*

A contractual relationship for construction services, labor, and materials between a project owner and a construction firm relies on a set of documents that is usually drafted by the project owner and their consultants. The main document is a contractual agreement, also known as the construction (prime) contract. This document defines the rights and obligations of each signatory party, including the construction firm. Drafted by the owners, prime contracts for construction services are submitted to the selected construction firm for review and acceptance. During the review, the parties may decide to re-negotiate some of the contractual terms before signing the contract. Various factors, including the nature of ownership (private versus public) as well as the project delivery and procurement methods, can have an effect if there is opportunity for negotiations and what can be negotiated. Private owners tend to provide more opportunities for negotiation than public owners do because public owners can rarely negotiate anything that may have affected the selection. For instance, if a construction firm was selected by a public agency based on their price and time offer, the contract sum and the delivery time could not be changed as that would disadvantage other bidders and most likely lead to claims. Most public procurement laws would forbid negotiating terms that were used to make the prime selection or affect the work post bid.

## *Subcontracts*

A contractual relationship between a construction firm and an entity other than the project owner also relies on a set of documents. The main document is a contractual agreement, also known as a construction subcontract. Similar to a prime contract, this document should define exactly and explicitly the rights and obligations of each signatory party. Subcontracts for construction services are usually developed by the buyer of the services, who is frequently a prime contractor, and replicate aspects of the prime contractual documents to share some of the prime contract risks with the subcontractor.

## *Pricing methods*

### *Cost plus a fee*

This pricing method, also referred to as simply cost plus, prices the contract by adding up two elements. The first element is the cost to perform the contract work, which is the sum of direct costs (costs of construction, including labor, materials, subcontracts, and equipment that can be directly related to specific work items) and the portion of indirect costs that are not directly related to any specific scope of work but apply to the entire contract. The second element is the fee, which can be expressed as a percentage of the cost element, as a fixed amount, or through an incentive fee. Under a cost plus with incentive fee, the contract "provides for the initially negotiated fee to be adjusted later by a formula based on the relationship of total allowable costs to total target costs" (GSA 2017; subpart 16.405-1). The fee element allows the construction firm to charge the owner for the expected profit and the portion of indirect costs that cannot specifically apply to the contract work. The indirect costs represent the cost of doing business — also known as home office indirect costs.

The cost plus method offers advantages and disadvantages to both parties. Since the pricing is cost-dependent, the payment method will require the construction firm to show its costs through open-book accounting. This provides transparent access to cost information to the purchasing party but requires more detailed recording and processing work as well as additions to the contract administration. Given their nature, cost plus contracts are especially advantageous when the design is not complete or for complex projects because they do not allow the contractor to hide contingencies. For instance, a project owner using this method to contract with a prime contractor would have information on all subcontracted work and prime contractor labor and material costs but would also need to review all invoices before processing a monthly payment. However, it also means that the contract price is unknown at the start, which creates an uncertainty about the contract price that cannot be used as a selection factor.

To address this price uncertainty, cost plus contracts can add a Guaranteed Maximum Price (GMP) condition. When a GMP is added to a cost plus contract, the contractor is taking the risk of cost overruns beyond the GMP. Under a cost plus with GMP contract, the owner is agreeing to pay for the work up to the GMP value, which acts as a price cap for the owner. However, it is difficult to secure a "firm" GMP too early in the design phase, as the contract work is still vaguely defined. If a cost plus contract is signed early on, for instance, under construction-management-at-risk (CMR) project delivery, the GMP cap would only be added later once the design package has evolved enough to allow the contractor to price it. As a rule of thumb, the design package is recommended to reach 60 to 90% of design completeness before an accurate GMP can be developed. These percentages should not be considered in absolute terms for each item of the scope of work, as discussed in the next example. There are circumstances where the owner and contractor may lock a GMP on specific scope items once they reach an adequate level of definition to be priced. To budget funds toward the remaining items that are still vaguely defined, it is customary for the parties to estimate their price at the current level of definition. These tentative prices are called allowances and are not intended to cap the price. Instead, they are used to guide design and to identify cost overruns early on.

### *Unit prices*

Unit pricing is a method that provides a ballpark contract price by summing estimated prices for each scope item. These prices are obtained by multiplying the quantities associated with each scope item — estimated by the project owner — by the unit prices, which are provided by the contractor. The use of the unit price method is common on public works and heavy civil projects, such as roads and utilities. Under this contract, the construction firm will provide a list of unit prices for various work items during procurement, such as cubic yards of excavation or the linear feet of a 6″ drainage pipe. Since these prices are dependent on quantities, the project owner provides estimated quantities and bears the risks associated with quantity estimating errors, whereas contractors bear the risks of pricing errors. This method is often associated with contractual adjustments for quantity variations outside a contractually set range. It is also often associated with a matching payment method where the project owner's inspectors evaluate quantities in place and compare them against billed quantities before authorizing progress and final payments. Since somewhat reliable quantities are necessary, design activities should be at an advanced stage at the contracting time similar to the lump-sum method. However, the unit price method does not transfer the risk of erroneous quantity takeoff to the contractor, which is beneficial for heavy civil projects where quantities may be only tentative.

### *Lump sum*

This pricing method, also referred to as stipulated sum or fixed sum, requires the general contractor to price the contract upfront. Therefore, this method is preferable for projects where the design is sufficiently complete so that the construction firm can reasonably estimate project costs, though it has also been used to price design–build contracts based on incomplete design packages. However, using this method when the design is incomplete tends to inflate the project price by incorporating contractor's contingencies for risks that may not materialize. While the lump sum price can cumulatively price all work to be performed, a schedule of values (SOV) is usually adopted to break the price down into work items and is used to identify progress payments through an assessment of the percentage of completion for each item. Still, SOVs can be used under any contract type. The CSI MasterFormat is a commonly used approach to present a schedule of values.

## Summary

The project delivery method is the owner's plan of how to deliver their project. Procurement and contracting define critical operating procedures for implementing that plan. Project delivery begins with the identification of the project objectives, and it continues with the selection of a project delivery method that would help fulfill these objectives. However, most project owners rely on other project participants for specific work items. Procuring and contracting these participants to the project team is similar to recruiting players for a football team. Once the team is assembled, the project will be executed with each team member playing a role on the field that is dictated by their contract. Selecting the wrong party for a role or providing that party with improper instructions through its contract may make it difficult for all parties to succeed on the project.

# About the Authors

**Pramen P. Shrestha** is a Professor of Construction Engineering Management and Graduate Coordinator in the Civil & Environmental Engineering and Construction department at the University of Nevada Las Vegas (UNLV). He holds an M.Sc. in Civil Engineering from Oklahoma State University and a Ph.D. in Civil Engineering from the University of Texas at Austin, United States (US). He completed his B.Sc. in Civil Engineering from the National Institute of Technology, Jamshedpur, India. He had about 10 years of professional experience in design and construction of highway and infrastructure projects in Nepal before moving to the US. After completing his Ph.D. in Civil Engineering from the University of Texas at Austin in 2007, he joined UNLV as an Assistant Professor. He has received federal, state, and private research grants worth about US $2.5 million. His research grants are funded by the US State Departments of Transportation, Transportation Research Board, National Science Foundation, US Labor Department, Water Design Build Council (WDBC), Airport Cooperative Research Project (ACRP), and private agencies. He has helped with the graduation of five Ph.D. and 34 master's students as an advisor. He currently has two Ph.D. students and three master's students working for him on various projects. He has published more than 55 peer-reviewed journal papers and 75 conference proceedings. He is a Fellow of American Society of Civil Engineers and a licensed professional engineer in Texas.

**Giovanni C. Migliaccio** holds a Howard S. Wright Endowed Professorship, he is the current Chair of the Department of Construction Management, and he has served as the Executive Director of the Center for Education and Research in Construction (CERC). He holds an M.S. and a Ph.D. in Civil Engineering from the University of Texas at Austin and a master-level degree from Politecnico di Bari in Italy. Prior to moving to the US, he worked in Italy in the construction management of telecommunication projects under Nortel Networks, Nokia Networks, and IPSE 2000. At UW, Giovanni is active in all three pillars of academic life, including teaching, research, and service to the university and the industry.

**G. Edward Gibson** is currently a Professor and holds the Sunstate Chair in Construction Management and Engineering in the School of Sustainable Engineering and the Built Environment (SSEBE) at Arizona State University. From 2010 to 2018, he served as the SSEBE School Director, overseeing significant growth in its programs and rankings. In addition to ASU, he served on the faculty of North Carolina State, University of Texas at Austin, University of Alabama, Tuscaloosa, and as an instructor at Auburn University. His educational background includes a B.S. and Ph.D. in Civil Engineering from Auburn University and an M.B.A. from the University of Dallas. Dr. Gibson has been PI or co-PI on over $11.0 million worth of funded research in his career with research and teaching interests that include front-end planning, safety leadership and systems, prevention through design, organizational change, asset management, alternative dispute resolution, knowledge management, earned value management systems, and risk management; he has received many awards for excellence in research, teaching, and service, including both the CII Outstanding Researcher (1996, 2004) and Outstanding Instructor (1998, 2014) twice. He is the developer of the Construction Industry Institute (CII) Project Definition Rating Index (the suite of five validated tools) and FEED

MATRS tools, has consulted with over 70 organizations, and taught over 215 short courses to industry. He has facilitated front-end planning risk assessment sessions on over 150 projects. He has over 240 publications and has given almost 300 presentations in his career. He is also an expert in educational leadership, having held a variety of successful academic management positions for 19 years during his career and consulted with over 20 universities. He has been active on many national committees, among them several National Research Council committees, Department of Energy Committees, the American Society of Civil Engineers (ASCE) Architectural Engineering Institute, and also served as a Fulbright Senior Specialist in Norway in Fall 2004. Dr. Gibson has several years of industry experience, has served as an Army officer, and is a licensed professional engineer in Texas. He was an elected member of the National Academy of Construction (NAC) in 2005 and a Distinguished Member of ASCE in 2020. He was awarded the 2016 ASCE R. L. Peurifoy Award for outstanding research and the 2020 Richard L. Tucker Service Award from NAC. He also served as a Visiting Academic Fellow at Cambridge University in spring 2019.

# Chapter 11

# Performance-Based Contracting Approaches

Kishor Shrestha

*Washington State University*

## Abstract

Today, agencies use several contracting approaches to outsource their work. When they contract out to private contractors, they must consider a few issues, including how the contractors perform their work, what the contractors' responsibilities are, what risks will be transferred to the contractors, and how the contractors will be paid. In other words, the agencies must determine what specifications they want to use for the contractors to perform their work. Based on specifications used in typical contracting approaches, two outsourcing methods are common in the construction industry: method-based contracting (MBC) and performance-based contracting (PBC).

Since the PBC method is comparatively a newer contracting approach, it warrants a more in-depth look. PBC use does not have a very long history, being first used in British Columbia, Canada, in 1988 to maintain roadway and bridge assets. The PBC method uses performance specification, which allows contractors freedom related to performing a given scope of work. More specifically, the performance specification does not bind the contractor, as does the traditional method-based specification, as far as 'how to do', 'when to do', and 'what to do' are concerned. The PBC approach also provides incentives and disincentives to the contractor based on the work quality produced.

Today, the PBC approach is being used globally. Agencies who have used this approach have benefited in various ways through saved cost, increased level of service (LOS), etc. Globally, transportation agencies have used the PBC approach for several types of work, such as roadway and bridge construction and maintenance, rest stop area operations, and mowing activities, to name a few. There is also a history of agencies implementing the PBC approach in building projects, including the service providing works of buildings in the operational phase, by reducing utility costs for example. Studies have shown that the main benefits of using the PBC approach are economic and managerial.

This chapter explains the reasons for using the PBC approach, types of PBC approaches, and PBC approach implementation in various project types, as well as case studies, their findings, and the challenges and lessons learned from past PBC contracts.

## Contracting approaches used in construction industry

Agencies use different contracting options to hire contractors for their projects. Based on the agencies' specifications, contracting approaches can be categorized into two main methods: traditional method-based contracting (MBC) and newer performance-based contracting (PBC).

### *Method-based contracting*

MBC is also known as an input-based contracting method or a prescriptive-based contracting method. This method has been used for a long time and is also called the traditional contracting method. It is still used extensively in various construction projects. In this type of contracting method, the project's submittals or requirements are explained using method-based specifications. With the named specifications, this method focuses on a project's input parameters and how various activities are done (Stankevich *et al.*, 2009). In other words, method-based contracting directs the contractor as to what they need to do, how they need to perform the work (how to install and test), when they need to perform the work, what and how much material to use (Stankevich *et al.*, 2009). One important reason for selecting this method is the availability of staff with specific skills (Shrestha *et al.*, 2016; Shrestha and Shrestha, 2014b). In MBC, payment to the contractor is based on the contractor's quantity of input or the amount of work completed (e.g., square yard of grading work or cubic yard of concrete pouring work is completed).

Let's discuss an example to present more information about method-based contracting. For highway rigid (concrete) pavement projects, state specifications and plans generally state what type of concrete will be used as well as the requirements to pour the concrete. During concrete pouring, QA/QC staff take samples to test. Below is a part of specification from Iowa DOT for their roads and bridges (Iowa DOT, 2015).

> *Proportion materials for pavement concrete in one of the mixtures identified in Materials I.M. 529 for the class of concrete specified in the contract documents. Any of the mixtures may be used, at the Contractor's option, provided the gradation of the separate aggregates complies with the gradation required for that mixture. Do not use C-5 and C-6 concrete mix proportions for pavements on Interstate or Primary highways. b. After October 31, use Class A concrete to construct all items of concrete pavement specified to be constructed with Class B concrete. The Engineer will either require completion by continuing placement operations past October 31, or allow the Contractor the option of a winter shutdown. When completion is required, the Contracting Authority will pay the net increase in cost of materials resulting from the change in proportions for any pavement placed within the contract period or authorized extension of the contract period. Other increases shall be at no additional cost to the Contracting Authority. (Iowa DOT, 2015, p. 264)*

The above excerpt from the Iowa DOT Specification explains how to do the concrete placing work. It also explains about the requirements for the concrete mixture, with part of the specification stating the concrete class to be used, along with how to place concrete after October 31.

With the MBC method, the chosen engineering firm takes responsibility for what the project will look like and how the contracting agency controls the contractor. The engineering firm prepares the specifications, which the contractor (project builder) will follow to complete the project.

This type of approach is suitable for the Design–Bid–Build (DBB) project delivery method.

Beyond the prescriptive approach, agencies also use the PBC approach for their projects. This approach is explained in the following section.

## *Performance-based contracting*

PBC is a relatively newer contracting approach, which is also called output-based contracting in some countries (Pakkala, 2002). It uses performance specifications, which explain the end requirements or performance targets of the works to be done. Unlike method-based specifications (explained in the previous section), performance specifications do not direct the contractors in what they need to do, how they need to perform the work (the procedure to do the work and test the work), or when they need to perform the work; however, the specifications clearly state the required performance of the PBC contractor. Various factors (saving cost, enforcing quality standards, improving LOS) influence selection of the PBC method (Shrestha and Shrestha, 2014a; McCullouch *et al.*, 2009). This approach rewards better work and penalizes worse work, which is the beauty of using the PBC approach.

Here is an example for a highway rigid pavement work.

> *The performance specification would require a minimum concrete compressive strength of 4,000 psi while using 1.5 — in dowels. The agency may also add more target performance such as the pavement shall function for certain number of years and a pavement condition index (PCI) shall not drop below 60 during the contract period. If the contractor performs better than target outcome (here compressive strength over 4,000 psi), a monetary incentive shall be provided; and if the contractor performance under the target outcome but within the acceptable tolerance limit (here compressive strength is in-between 3,800–4,000 psi, a monetary penalty shall be charged to the contractor. And. if the contractor's work performance is under the acceptable tolerance limit (here compressive strength below 3,800 psi), the work shall not be accepted.*

Payment to the PBC contractor is tied to the work performance target. If the contractor fulfills the performance targets (set in the specifications), the contractor is paid on a regular basis, usually spread in equal amounts every month for the duration of the contract. The engineering firm (with the contracting agency) prepares the performance specifications to set out the end requirements, or performance targets, of the project, and the contractor works to achieve them. The performance targets may come with a range of acceptance based on timeliness. Timeliness is the standard time duration (specified in a contract) provided to the contractor to fix any defects on their work during the contract period. It varies from agency to agency and project to project as per the nature or works. If the contractor performs better than the target performance, the contractor receives an incentive (this may not happen in all projects), and if the contractor performs below the target performance, a disincentive may be charged to the contractor (Schexnayder and Ohrn, 1997; Popescu and Monismith, 2006). The incentive and disincentive amounts are decided on a case-by-case basis. Most agencies provide both incentives and disincentives, but some agencies provide just one, either incentives or disincentives.

With the PBC approach, the contract duration commonly refers to the time duration in which the contractor is responsible for construction and/ or operation and maintenance (O&M). The contractor is fully responsible for maintaining the asset(s) at or above the minimum acceptable condition, as stated in the specifications. For example, the Nevada DOT PBC pavement surface striping project had a five-year contract duration, in which the PBC contractor was responsible for maintaining a good visibility of the pavement striping lines for five years.

In PBC, the contract duration is important to consider. The contract duration depends upon various factors, with most agencies starting with a shorter duration (typically three to five years). The minimum duration for a chip seal contract is three years (Shrestha, 2016). However, as agencies gain experience with the PBC approach, they go with longer contracts (over 10 years). For example, the initial PBC contract in Canada was three to five years, and gradually the contract was extended up to 10 years (ADB, 2018). The average contract duration with the PBC approach has been calculated as 6.8 years (Pakkala, 2005). Studies show longer contract durations saved more in expenses for agencies, and most economically successful contracts have a long-term duration (over five years); however, longer contract durations produce more risk.

For the selection of a PBC contractor, agencies use either the 'best-value method' or the 'qualifications-based method'. The best-value method uses a combination of two parameters: the contractor's bid price and technical qualifications. As this method takes into consideration the contractor's price on a project as well as their qualifications, it offers the best value to contracting agencies. The weights of these two parameters (price and qualification) vary contract to contract and state to state with project type. Some agencies use a criterion of 75% price and 25% qualifications, whereas other agencies used 50% price and 50% qualifications (Pakkala, 2002).

PBC contracts in British Columbia, Canada have used 40% weight on price and 60% on qualifications; PBCs in Alberta, Canada have used 78% price and 22% qualifications; Finland's PBC contracts have used 75% price and 25% qualifications; Sweden's PBC contracts have used 90% price and 10% qualifications (NCHRP, 2009; Pakkala, 2002).

The 'qualifications-based method' uses the qualifications of the bidders in selecting a contractor. This method focuses on quality and value. Price is not a factor for bidder selection in this method; however, parties will negotiate the price of completing work prior to signing a contract (AGC, 2009). In Australia, the qualifications-based method has been used to select a bid winner for a 'Pure PBC' contract (Wirahadikusumaha *et al.*, 2015). The 'Pure PBC' approach is explained in Section 10.4 (types of PBC contracts).

Figure 1 illustrates how the PBC approach maintains an asset/infrastructure. Let's take an example of chip seal road section. The red dash line represents the minimum level of service (LOS) of the chip seal pavement to be maintained, which is generally mentioned in the specification. One of the performance measures used for chip seal pavement is International Roughness Index (IRI). It indicates driving comfort level and safety; thus, IRI is an important parameter in calculating PCI value of a road pavement. For this example, let's assume the specification states the PCI of the chip seal pavement shall exceed 60. Right after placing the new chip seals on the existing pavement surface, the PCI value will be above 90 and the LOS of the pavement condition will be at the highest level, which is indicated by "A" point in the Fig. 1. As the pavement

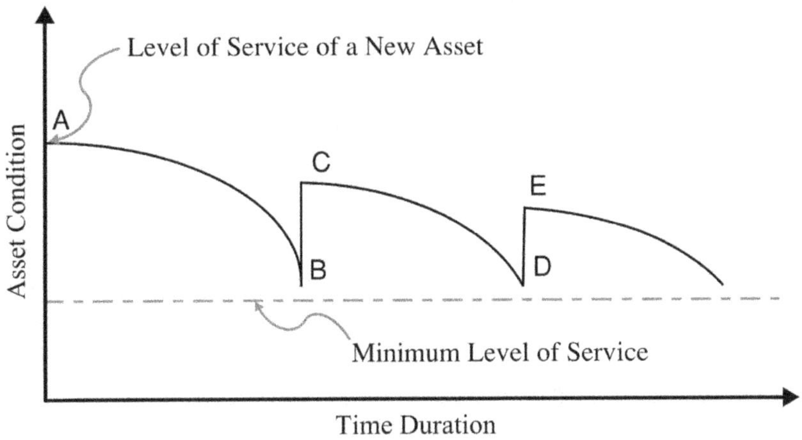


**Figure 1:** Asset condition over time duration using the PBC approach.

surface being used over the years, the LOS decreases as the PCI decreases; however, the pavement surface does not need any maintenance until the PCI is 60. As the PCI approaches close to 60, which is indicated by "B" point in the Fig. 1, the PBC contractor may need to apply a new chip seal or other strategies to bring the PCI back to above 60 again, which is indicated by "C" point in the Fig. 1.

## Types of performance-based contracting

### *Types based on performance measurement*

Basically, PBCs are categorized into two types based on how the contractor's work performance is measured. They are 'Pure PBC' and 'Hybrid PBC' approaches. With the Pure PBC approach, as task outcomes are completed, they are measured using the PBC approach with the performance indicators. Generally, the number of activities is limited with this approach. An Australian 'Pure PBC' was used for routine and periodic maintenance, emergency response, reconstruction and rehabilitation, collection and maintenance of asset inventory data, and customer relations activities of roadways (Wirahadikusumaha *et al.*, 2015). As the 'Pure PBC' focuses on output quality, this approach uses a qualifications-based method for contractor selection.

The Hybrid PBC is a combination of two contracting methods. Commonly, the 'Pure PBC' is combined with a traditional contracting approach, using the method-based specification. With this method, some activities are measured based on 'Pure PBC' strategies, and some activities are measured based on method-based specifications, which are input-based. One of the reasons contracting agencies use the 'Hybrid PBC' approach is due to the number of activities that can be bundled into a single contract. Activities that need higher quality output and customer satisfaction are key and can be combined with other common roadway activities in this approach. This 'Hybrid PBC' approach can be cost-efficient, as selective activity outputs are measured based on the contractor's performance, which provides higher road user satisfaction, whereas other activities can be measured based on a traditional contracting method. In an Australian contract, a number of urban and rural road maintenance activities were combined for a 'Hybrid PBC' (Wirahadikusumaha *et al.*, 2015).

## *Types based on assets included*

Based on the nature and types of roadway assets included in a contract, PBC contracts can be categorized into nine types. Per NCHRP (2009), the nine types of the PBC contracts for transportation asset maintenance include the following:

1. *Single activity contract*: These PBC contracts include a single activity, for example, a PBC striping contract in Nevada that includes striping on Interstate 95.
2. *Single asset contract*: In this type of contract, all of the activities under a specific asset type are included in a single contract. The best example of this type of contract is bridge asset maintenance.
3. *Set of related activities contract*: This type of contract includes a set of activities that are related to each other. Rest area maintenance is an example of this type of contract.
4. *Fence-to-fence contract*: This is also termed as a corridor contract. All of the activities inside the right-of-way of a highway are included in

a fence-to-fence contract. Washington D.C., Virginia, and Florida have used this type of contract.

5. *Area-wide contract*: In this type of contract, an activity or an asset in an area (city, county, district, or state) is included in a single contract. Montana DOT awarded a PBC contract to maintain a number of their safety rest areas, but not all safety rest areas, in this type of contract.
6. *Hybrid contract*: There can be various types of hybrid contracts. Basically, this contract includes traditional input-based specifications and performance-based specifications.
7. *Agency-to-agency contract*: With an agency-to-agency contract, basically, the public agencies in a state, county, city (or other agencies) are responsible for carrying out the state's projects. Generally, roadway maintenance work is contracted out with this type of contract. In Michigan, a number of counties perform this type of contract.
8. *Warranty-based contract*: When agencies outsource maintenance work to contractors with a warranty, warranty-based contracts are used. With this kind of contract, the contractor will be responsible for fixing any defects (e.g., LOS of pavement surface) with their work for a specified period of time. This contract may be useful for certain types of road maintenance work, such as rest areas, signs, and pavement surfaces.
9. *Multi-phase contract*: When a contract needs to cover multiple phases of an activity's life cycle, a multi-phase contract can be used to maintain the asset. For example, the design–build–operate–maintain (DBOM) project delivery method is widely used globally. With the DBOM delivery system, the contractor (or joint venture with other firms) may maintain the assets constructed under the PBC approach.

## *The execution of PBC in various projects*

The history of the PBC approach is not very long. However, this contracting approach has been implemented for several types of projects since its first application in transportation asset management work in British Columbia, Canada in 1988. This section discusses the use the PBCs in various types of projects.

The PBC method is extensively used in horizontal projects, such as new construction and roadway and bridge asset maintenance. Since the PBC was introduced in the United States, this approach has been used in at least 16 states on urban roadway and interstate highway projects. Michigan DOT used PBC for five miles of new road construction and to replace two bridges (USDOT, 2013). Virginia, Florida, Idaho, Maryland, New Mexico, Texas, Utah, Montana, and Washington D.C. have also used this approach to maintain a number of highway assets (NCHRP, 2009; Shrestha and Powers, 2018). Texas and Montana used this approach to maintain their safety rest areas along highways. Oklahoma and Washington D.C. used PBC to maintain bridge assets. Almost all of these contracts used lump-sum fixed price contracts, with provisions of incentives and disincentives to the contractors.

PBC contracting has also been used in various civil engineering infrastructure projects. One project was urban water supply and sanitation in Eastern Europe (OECD, 2006). The result-oriented contracting strategy was used to improve both the physical and financial issues of the project. To make this project successful, several feasible performance indicators were developed, which could be measured to assess the contractor's performance.

Agencies have been using the PBC approach for service providing projects in buildings for a long time. PBC has been used for commercial, government, and industrial buildings and is common for energy service providing contracts. In these contracts, contractors provide services under the PBC approach. In PBC transportation projects, mobilization (certain percentage of the total contract value) is commonly provided to the contractors. However, with the service providing projects, mobilization is not provided to the contractor. Two examples of using the PBC approach for utility service providers in buildings are explained below.

*The government entities and institutions has implemented the PBC method to reduce the utility expenses and operational costs. To be noted that there was an energy crisis in the early 1970's. For these works as well as to install any equipment, the contractor was not paid up-front. In Redlands city, California, the city implemented PBC contract for capital improvement in order to reduce the utility cost of the buildings. With that contract, the utility cost was reduced to half in the two years of service and the contractor was paid from that saving funds. Similarly, a number of entities at the state and local level benefited from the PBC services to reduce the energy expenses in Tennessee.* (Gordon 1998)

*Similarly, with the "Green Deal" project, a contractor was hired to provide energy saving solutions to the UK Department of Energy and Climate Change. The contractor supplied all the required equipment. The major objective of this project was to reduce the energy consumption in homes using various types of equipment including wind turbines and solar panels, installing insulation, using double or triple glazed windows and other suitable techniques in the homes. The contractor's performance was measured on the amount of energy reduced in the homes. The contractor was paid as a part of savings in the energy expenses in that homes.* (UR, 2013)

## The global perspective of performance-based contracting

Implementation of the PBC approach in transportation construction/maintenance projects began in 1988 in British Columbia, Canada to maintain a large volume of roadway and bridge assets (Zietlow, 2004). This approach was also introduced in Australia, New Zealand, England, and Finland. In the United States, it was used for the first time in 1996 for a transportation project (Zietlow, 2004). Many countries in the world (America, Europe, Asia, and Africa) are taking advantage of using the PBC approach for new construction, as well as to manage and maintain infrastructure, and for performance services. This approach has been dominantly used in the transportation sector. Table 1 presents the countries where this approach has been used.

### *North America*

Canada was the first country to outsource their transportation projects using the PBC approach in the late 1980s (Zietlow, 2008). After the first PBC project was successful, this approach was replicated in various provinces around Canada as well as in the United States and other countries, to maintain transportation assets.

**Table 1:** Countries who implemented (and plan to implement) the PBC approach.

| S.N. | Sub-continents | Countries | Types of project | Data sources |
|---|---|---|---|---|
| 1. | North America | USA | Road and bridge assets | (ADB, 2018; Pakkala, 2005; Zietlow, 2008) |
| | | Canada | Road and bridge assets | (NCHRP, 2009; ADB, 2018; Pakkala, 2005; Zietlow, 2008) |
| 2. | Central America | Honduras | Roadway assets | (Zietlow, 2008) |
| | | Guatemala | Roadway assets | (Zietlow, 2008) |
| | | Nicaragua | Roadway assets | (Zietlow, 2008) |
| 3. | South America | Argentina | Roadway assets | (NCHRP, 2009; ADB, 2018; Zietlow, 2008) |
| | | Brazil | Roadway assets | (Gericke, 2014; Zietlow, 2008) |
| | | Chile | Roadway assets | (Zietlow, 2008) |
| | | Columbia | Roadway assets | (Zietlow, 2008; Stankevich, 2009) |
| | | Peru | Roadway assets | (Zietlow, 2008) |
| | | Uruguay | Roadway assets | (NCHRP, 2009; ADB, 2018; Zietlow, 2008) |
| 4. | Australia/ New Zealand | Australia | Roadway assets | (NCHRP, 2009; Pakkala, 2005; Zietlow, 2008; Wirahadikusumaha *et al.*, 2015) |
| | | New Zealand | Roadway assets | (NCHRP, 2009; ADB, 2018; Pakkala, 2005; Zietlow, 2008) |
| 5. | Northern Europe | Denmark | Roadway assets | (ADB, 2018) |
| | | Estonia | Roadway assets | (ADB, 2018; Pakkala, 2005) |
| | | Finland | Roadway assets | (NCHRP, 2009; Pakkala, 2005) |
| | | Norway | Roadway assets | (Pakkala, 2005) |
| | | Sweden | Roadway assets | (Pakkala, 2002; 2005) |
| | | United Kingdom | Roadway assets | (NCHRP, 2009) |
| 6. | Western Europe | France | Roadway assets | (Stankevich, 2009) |

**Table 1:** (*Continued*)

| S.N. | Sub-continents | Countries | Types of project | Data sources |
|---|---|---|---|---|
| 21. | Eastern Europe | Poland | Roadway assets | (Gericke, 2014) |
| 22. | Southern Europe | Albania | Planned to implement for roadway assets | (Stankevich, 2009) |
| 23. | | Serbia | Roadway assets | (ADB, 2018) |
| 24. | | Montenegro | Roadway assets | (ADB, 2018) |
| 25. | Western Asia | Azerbaijan | Planned for state roads of 610 miles (982 km) beginning from 2018 | (ADB, 2018) |
| 26. | | Georgia | Roadway assets | (ADB, 2018) |
| 27. | | Yemen | Planned to implement for roadway assets | (Stankevich, 2009) |
| 28. | Central Asia | Afghanistan | Roadway assets | (ADB, 2018) |
| 29. | | Kyrgyzstan | Planned to implement for roadway assets | (ADB, 2018) |
| 30. | | Tajikistan | Roadway assets | (ADB, 2018) |
| 31. | East Asia | Mongolia | Planned to implement for roadway assets | (ADB, 2018) |
| 32. | South Asia | India | Roadway assets | (Gericke, 2014; Stankevich, 2009) |
| 33. | | Pakistan | Roadway assets | (ADB, 2018) |
| 34. | | Nepal | Roadway assets | |
| 35. | Southeast Asia | Cambodia | Planned to implement for roadway assets | (ADB, 2018; Stankevich, 2009) |
| 36. | | Indonesia | Roadway assets | (ADB, 2018; Stankevich, 2009; Wirahadikusumaha *et al.*, 2015) |
| 37. | | Malaysia | Roadway assets | (ADB, 2018) |

(*Continued*)

**Table 1:** (*Continued*)

| S.N. | Sub-continents | Countries | Types of project | Data sources |
|---|---|---|---|---|
| 38. | | Philippines | Roadway assets | (ADB, 2018; Gericke, 2014) |
| 39. | | Thailand | Planned to implement for roadway assets | (Gericke, 2014; Stankevich, 2009) |
| 40. | | Vietnam | Planned to implement for roadway assets | (Stankevich, 2009) |
| 41. | North Africa | Egypt | Roadway assets | (Gericke, 2014) |
| 42. | West Africa | Burkina Faso | Planned to implement for roadway assets | (Stankevich, 2009) |
| 43. | | Cape Verde | Planned to implement for roadway assets | (Stankevich, 2009) |
| 44. | | Liberia | Roadway assets | (Gericke, 2014) |
| 45. | | Nigeria | Roadway assets | (Gericke, 2014) |
| 46. | Central Africa | Chad | Roadway assets | (Stankevich, 2009; NCHRP, 2009) |
| 47. | East Africa | Madagascar | Planned to implement for roadway assets | (ADB, 2018; Stankevich, 2009) |
| 48. | | Tanzania | Roadway assets | (Gericke, 2014; Stankevich, 2009) |
| 49. | | Zambia | Roadway assets | (Gericke, 2014; Stankevich, 2009) |
| 50. | Southern Africa | South Africa | Roadway assets | (ADB, 2018; Stankevich, 2009) |

In the United States, the PBC approach has been used by at least 17 state DOTs. PBC was a comparatively newer method, and many contract experiences were positive. The states who have used or are currently using this approach are Florida, Virginia, Texas, Oklahoma, Montana,

Nevada, Indiana, Michigan, Missouri, Nebraska, North Carolina, Pennsylvania, Idaho, Maryland, New Mexico, Utah, and California (NCHRP, 2009; Shrestha and Powers, 2017; SACHS, 2004; USDOT, 2013). The three states that have used this approach on larger scales are Florida, Virginia, and Texas. The PBC approach was also used to maintain 75 miles of the National Highway System in the District of Columbia (NCHRP, 2009). The types of project for which the PBC approach has been used were fence-to-fence projects and asset management projects as well as individual asset management projects to maintain chip seals, striping, mowing, safety rest area operations, etc. The cost savings with using the PBC approach have ranged from 10% to 15% (Pakkala, 2005).

## *Central and South America*

Argentina is one country where a large national roadway system has been maintained using the PBC approach. The Argentine government has moved from using the traditional MBC approach to newer PBC approach. The key reasons for moving to the PBC approach were to save cost, transfer project risks to contractors, foster innovation, seek out long-term funding, and increase road users' satisfaction levels (NCHRP, 2009).

A national road inventory survey was carried out in Argentina at the start of the rehabilitation and maintenance contracts (NCHRP, 2009). In that survey, a base line of the existing road network was prepared, which presented the existing traffic conditions and road conditions. Based on the traffic volume, roads were categorized into two groups: traffic over 3,000 vehicles per day and in between 300–3,000 vehicles per day (NCHRP, 2009). The contract was originated in 1995 for over 2,200 miles (over 3,540 km) of routine maintenance (NCHRP, 2009). When the PBC contract was reviewed and found as satisfactory, this approach was expanded. By 2009, 44% of Argentina's roadway assets ware maintained using the PBC method (NCHRP, 2009).

At the start of the contract (in Argentina), 5% of the total contract, followed by two 10% allotments, were dispensed to the contractor as mobilization (NCHRP, 2009). The remaining contract funds were divided into equal monthly installments over the duration of the four-year contract. The monthly payments were issued to the contractor based on the achieved target performance. To identify whether the contractor achieved target performance, the work was assessed every month. The monthly inspections were carried out by the contracting agency's engineers with the contractor present. The assessment was conducted based on four performance indicators: roughness, rut depth, cracking, and raveling. As a summary of the contract, the cost overrun and delay to work were nil. This project was also successful through cost savings and transferring project risk to the contractor. The performance indicators of this project were easy to measure and monitor. Once the initial contract for the project was deemed successful, a new contract was made with the same party for four more years.

After Argentina's success using the PBC method for their roadway maintenance, this method was replicated in other neighboring countries. After Argentina implemented the PBC method, Uruguay followed, with later implementation in Brazil, Columbia, Chile, Peru, Ecuador, Mexico, and Guatemala (Stankevich *et al.*, 2005).

## *Australia and New Zealand*

In 1995, Australia introduced output-based contract for the first time on an urban roadway system covering over 280 miles (over 450 km) in Sydney (Frost and Lithgow, 1996; Stankevich *et al.*, 2009; Zietlow, 2004). After achieving success with this contract, several additional transportation projects adopted this approach in some form, in both the southern and western parts of Australia, to manage the broader classification of roads from urban-to-rural (Wirahadikusumah *et al.*, 2015). The Australian government used both PBC approach types ('Pure PBC' and 'Hybrid PBC') to maintain their roadways. The 'Pure PBC' and the 'Hybrid PBC' are explained in an earlier section (Types of Performance-based Contracting).

Short-term pilot projects of over 62 miles (100 km length of roadway) were used first to learn the outcomes of the PBC approach before stepping up to outsource large road networks. To select the PBC contractor, the best-value contractor selection method was used, considering both the contractors' prices and technical qualifications. After achieving a successful outcome from the pilot projects (16% cost savings and 22% increased productivity), other regions replicated the PBC approach to manage and maintain several roadway systems (Wirahadikusumah *et al.*, 2015). In Sydney, New South Wales PBC project, they achieved 38% cost savings, with 30% in a Mid North Region project, and 25% cost reduction in a South Perth project (Frost, 2001). Cost savings using the PBC method in Australia have ranged from 10% to 40% (Pakkala, 2005), with the scope of work including maintenance, rehabilitation, and capital improvement. In Western Australia, the government used the PBC approach for 8- or 10-year contracts to maintain national, state, and local road networks, ranging 970–2,660 miles (1,560–4,280 km) (Wirahadikusumah *et al.*, 2015). In these contracts, a government superintendent was responsible for identifying any defects with the contractors' work and for issuing rework orders.

In New Zealand, the PBC approach was executed for the first time in 1998 (Zietlow, 2004; Stankevich *et al.*, 2009) for a project that covered over 250 miles (400 km) of national roadway maintenance (Stankevich *et al.*, 2009). This was a long-term contract and was called Performance-specified Maintenance Contract (PSMC) (NCHRP, 2009). At the end of the contract, it was declared that the PSMC approach delivered increased LOS of the roadways as well achieved significant cost savings (NCHRP, 2009; Stankevich *et al.*, 2009). After the first PSMC approach was successful, it was adopted in other areas and has since been implemented for maintaining transportation assets. In New Zealand, the main reasons for moving from the traditional contracting method to the newer performance approach were to achieve the key benefits of using the PBC approach. In fact, it was reported that the cost savings with using the PBC method was 20–30% (Pakkala, 2005). Both Australia and New Zealand have noticeably used the PBC method to successfully complete road maintenance projects (Stankevich *et al.*, 2005, 2006).

### *In other countries*

The PBC approach has been adopted globally, and Table 1 presents at least 39 countries which have implemented, along with 11 countries who plan to implement, the PBC approach for construction and/or maintenance of roadway and bridge assets. The PBC method has been remarkably successful in the countries/regions presented above, and this method has also been increasingly used in other countries such as Sweden, Netherlands, Norway, France, Estonia, Serbia and Montenegro, South Africa, Zambia, Chad, and Philippines (Stankevich *et al.*, 2005). One hundred percent of national roads in South Africa and 63% of national roads in Estonia are maintained using the PBC method. Similarly, in Chad, 17% of all season roadways are maintained using the PBC method. PBC method has also been implemented in several other countries such as Afghanistan, Denmark, Egypt, England, France, Finland, Georgia, Guatemala, Holland, Honduras, India, Indonesia, Liberia, Malaysia, Nicaragua, Nigeria, Norway, Pakistan, Peru, Philippines, Poland, Serbia and Montenegro, Tajikistan, Tanzania, and Zambia (ADB, 2018; NCHRP, 2009; Zietlow, 2008; Gericke, 2014; Pakkala, 2005; Stankevich, 2009). Furthermore, several countries plan to implement the PBC method including Albania, Cape Verde, Madagascar, Burkina Faso, Tanzania, India, Kyrgyzstan, Cambodia, Vietnam, Yemen, etc. (Stanakevich, 2006).

## *Legal aspect of the PBC approach*

A construction contract is defined as a lawful mutual agreement between a contracting agency, owner, and contractor. The contracting agency prepares the contract documents, which explain the work scope, project cost, project duration, parties' responsibilities, etc. If the state is executing a PBC project for the first time, while preparing the contract document, the agency may need to make sure it does not violate state laws and/or regulations. Per the Legal Information Institute (under CFR 37.101), a PBC is defined as:

> *Performance-based contracting means structuring all aspects of an acquisition around the purpose of the work to be performed as opposed to either the manner by which the work is to be performed or broad and imprecise statements of work.* (LII, 2021)

Unlike most traditional contracts, the contract durations of PBC projects are relatively longer (commonly three to ten years). If a contracting agency is using a new approach like the PBC, it would be helpful for the state contract teams to ensure they are following both state and federal law. This is why states include legal officers on their project teams, particularly during the planning and procurement phases.

Construction disputes are common in construction projects (Hinze, 2011; Shrestha *et al.*, 2021). Therefore, it is important for a project team to be knowledgeable about construction disputes and the methods of dispute resolution. Here is an example from a study conducted by Ribreau (2003):

> *Over a 2,500 lane mile highway maintenance contract (routine maintenance including snow and ice removal) was awarded to a PBC contractor for five years. In a few months of the contract awarded, the state started withhold payment to the contractor due to the contractor's performance issues. Over a six-inch snowstorm hit the project area in March 2002 that resulted severe public criticism of the contractor's performance. Ultimately, the contract was cancelled and they went in litigation.* (Ribreau, 2003)

## Why agencies choose PBC: Benefits to agencies and contractors

There are various reasons for choosing the PBC approach for construction projects. Figure 2 presents the primary benefits for an agency and contractor, which are discussed in this section. The benefits can be categorized into the following five items.

### *Economic benefit (cost saving)*

Economic benefits are a key reason for using the PBC approach. Because of budget limitations, contracting agencies (local, state, and federal) look for ways to reduce project costs without compromising LOS (output quality) (Shrestha and Shrestha, 2016).

Comparing and calculating the true economic benefits from a project is a challenge. Some projects may have higher upfront costs and lower operation and maintenance costs. Other projects may have just the opposite. Thus, comparing these projects and calculating the true

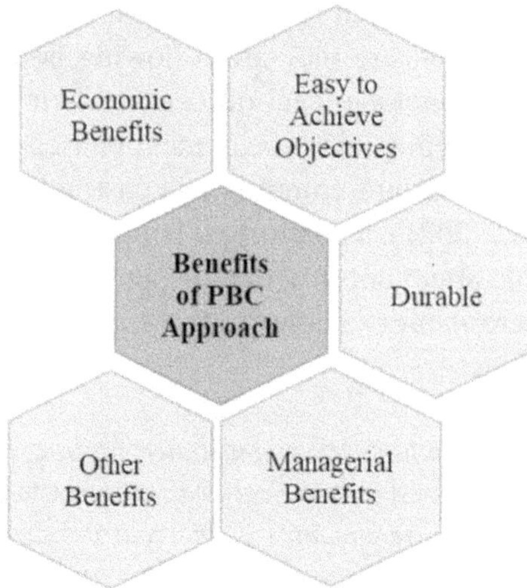


**Figure 2:** Benefits of using the PBC approach.

economic benefits, costs, or cost savings is not easy. To compare the alternatives, in order to calculate true cost savings, a life-cycle cost analysis can be carried out, which finds the most economical alternative (Shrestha *et al.*, 2019; Shrestha and Shrestha, 2013). For example, concrete pavements are expensive upfront, as compared to flexible pavements (asphalt pavement), but as the life of the rigid pavement is longer and has fewer operational and maintenance expenses, if the life-cycle costs of concrete pavement and flexible pavement are calculated in terms of the total cost per year, the dollar cost per year of rigid pavement may be less. Therefore, when calculating the economic benefits or cost savings for using the PBC approach, it is important to compute the life-cycle costs of the assets maintained by the PBC as well as other approaches.

Contracting agencies have moved to using the PBC method from various other methods. Some agencies shifted from using an in-house workforce approach, whereas other agencies moved from using a traditional contracting approach. Therefore, when computing the cost savings, it is also important to know what the agencies were spending before using the PBC approach. The life-cycle costs of construction and/or maintenance of both alternatives should be calculated and compared in order to estimate the true cost savings using the PBC approach. The PBC approach could also bring additional benefits, such as environmental (reduced carbon footprint or less pollution in nearby water bodies) and social benefits. To compute the true costs saved using the PBC approach, these additional benefits also need to be considered and converted into a dollar amount, which is challenging.

Additionally, long-term contracts offer more cost savings. PBC contracts are commonly longer than other contracts, such as traditional method-based contracting. The majority of PBC contracts are between three and ten years, but some contracts can be longer than 15 years.

> *The first PBC contract in Canada for a road and bridge maintenance project was awarded for 15 years.* (NCHRP, 2009; Zietlow, 2008)

When the contract period is longer, contractors focus on the quality of the work output. This is because quality work lasts longer. The high-quality work also decreases the life-cycle cost of an asset, which offers higher profit to the PBC contractors. PBC contractors also bring innovative approaches to make the work easier for greater economic benefits. A PBC contract reduces both direct and indirect work costs (Straub, 2009). Studies showed that using the PBC method achieved cost savings up to 40% (NCHRP, 2009; Zietlow, 2004). One of the required preconditions to be successful in achieving cost saving is that the contract period should be long (over three years) (Shrestha, 2016), so that the contractor will have the opportunity to plan and design in their own way toward a lower total life-cycle project cost.

## *Easy to achieve project objectives*

Other reasons that agencies select the PBC approach are the ease of meeting project objectives and getting work done to their satisfaction. That means that since PBC is an outcome-based contracting approach, agencies set the target outcome of the product. In other words, the agencies specify the expectations of the end product in the contract document.

For example, a state DOT is hiring a PBC contractor for concreting a rigid pavement; the state DOT may fix the properties of the concrete as they need to, such as they may clearly specify that the compressive strength of concrete shall be 6,000 psi or more when tested on the 28th day of pouring. Therefore, with using the PBC method, the outcome quality or end product of a project is easily controlled by the agency upfront, before the contract is awarded to the PBC contractor.

With most engineering projects in a traditional outsourcing approach, there is an acceptable tolerance limit of the end work. The PBC contracting also provides that privilege to the contractor. The acceptable tolerance limit is the range within which the contractor's work is still acceptable; the range may be 5 to 10% below the target outcome (stated in the specification), but can greatly vary depending on the state and nature of the work.

Some agencies may offer a greater amount of tolerance limit, whereas others may not. For instance, the target outcome of concrete's compressive strength is 6,000 psi, but the acceptable tolerance limit may be set as a range of 5,800 psi to 6,000 psi. Commonly, with a PBC, when the work output is below the target outcome but within the acceptable tolerance limit, there may be a disincentive or penalty to the contractor (contractor's monthly payment may be reduced until the defect is fixed). If the contractor's work output is below the acceptable tolerance limit (let's say the concrete's compressive strength is 5,500 psi), the work is rejected, and the contractor needs to do the work again without additional payment. On the other side, if the PBC contractor performs above the target, the contracting agency may reward the contractor. This is the beauty of using the PBC method, that is, rewards for good work and penalties for poor work. This is common with PBC projects.

The nature of rewarding a PBC contractor is generally through monetary incentives. The amount of incentives also varies greatly agency to agency and project to project, based on the varying nature of the work. With state highway maintenance contracts, some agencies have provided incentives up to 5%, whereas other agencies have provided incentives up to 30% (Popescu and Monismith, 2006).

> *For a new construction asphalt concrete pavement project in 2005 in California, two performance models were developed for two common*

*pavement distresses of rutting and fatigue. Per contract document, the contracting agencies shall provide monetary reward for the contractor's work above target performance and shall charge penalty for the work below the target. Caltrans provides a maximum of five percent monetary incentive for superior works.* (Popescu and Monismith, 2006)

## *Durable*

As explained in the 'Economic Benefit' section, a PBC contractor's projects last longer, and the products PBC contractor's work on also last longer (Shrestha, 2016; Shrestha *et al.*, 2017; Gericke *et al.*, 2014). Since PBC contractors are responsible for fixing any defects in their work during the entire contract period, agencies can expect a high-quality work output, which produces durable work. Moreover, PBC contractors are not bound by specifications on how to do the work, which fosters innovation, which in turn maximizes product durability. Some PBC specifications clearly state the contractor's work shall be durable for a certain time period (Hartwig *et al.*, 2005). In fact, one of the reasons that state DOTs have selected the PBC approach is because the work completed lasts longer (Shrestha, 2006).

## *Managerial benefits*

Another key benefit of using the PBC approach is that it reduces an agency's managerial strain. One managerial strain in construction projects is due to the risks involved with projects (Shrestha, 2018). However, adopting the PBC approach transfers the agency's risk to the contractor. When agencies perform construction projects internally or outsource the contracts through traditional contracting approaches, they face typical risks such as labor shortages, unknown site conditions, including geotechnical issues, managing change orders, safety hazards, incomplete plans and drawings, poorly written contract documents, soaring material and labor prices, vandalism, scheduling errors, project team conflicts, and failure to comply with quality requirements.

When the work is outsourced to a traditional contractor, the limited risks mentioned above are transferred to the contractor, such as safety hazards and soaring material and labor prices, but the majority of the risk will remain with the contracting agency. When contracts are outsourced through the PBC approach, most risks can be transferred to the contractor. When a contracting agency's risks are transferred to a PBC contractor, it lessens the agency's managerial strain (Shrestha and Powers, 2018). It also should be noted that the amount of risk transferred to the contractor will impact the cost of the project; the more risk transferred to the contractor, the more project cost goes up, and vice versa. Thus, in most projects, risks are carefully shared between the contracting agency and the contractor.

Project monitoring and supervision is another strain on the management team. Often, a PBC project mandates the use of a third party to monitor and supervise the contractor's work and reports back to the agency. With some projects, monitoring is done by both a third party and high-level agency administration. This helps to reduce the contracting agency's strain on managing the staff for monitoring and supervision.

With the PBC approach, several assets can be bundled into a single contract. Bundling assets reduces the number of contracts with an agency, which ultimately reduces the administrative burden. The PBC contractor is not paid with a detailed measurement of all the tasks, which also reduces the administrative effort. Additionally, cost overrun is prevalent in almost every project outsourced without using the PBC method (with some exception). Project delays are another issue, especially in large complex projects, when outsourced through the traditional contracting methods. Cost overrun and project delays place strain on the project management team, which they always try to avoid. The PBC approach provides a good alternative for agencies to mitigate these issues.

Moreover, the PBC method allows for bundling a number of construction and maintenance activities into a single contract. One example is the first PBC contract in British Columbia, Canada in 1988 (Zietlow, 2004). This contract included a number of roadway and bridge asset maintenance activities. Bundling a number of projects in a single contract significantly reduces the number of contracts with an agency. This saves time and money preparing the contract documents and tracking them, as this is a very cumbersome exercise for the project management team to undertake. Therefore, bundling the projects and making a larger contract may also help an agency to reduce their managerial stress.

In addition, a PBC contract helps management to secure funding for a longer time period. As the PBC contract duration commonly varies from project to project and agency to agency, from three to ten years, before planning a PBC contract, the agency needs to make sure continuous funding will be available for a minimum of the contract duration. As the long-term funding is secured for a project at beginning of the project planning, the project team will not need to prepare the contract document every year, leaving them free of stress for budgeting and potentially reducing the number of employees monitoring the contract as well. For these reasons, agency managers' preference for selecting a contracting method for new construction and maintenance projects is the PBC method in several states, proven by the increasing number of PBC contracts initiated every year.

## *Other benefits*

Using PBC has additional benefits. This contracting approach provides freedom to the contractor, which fosters innovation. There are several examples that demonstrate how the PBC approach brought about innovation. The graffiti reduction using a CCTV camera is one example. Other benefits include the following:

- The PBC approach provides transparency to the public. The PBC contractor provides information regarding product conditions (road conditions) (McCullouch *et al.*, 2009).
- The PBC approach delivers higher customer satisfaction.
- With the PBC contracting approach, the contracting agency may provide automatic extensions of the contract term if the PBC contractors do a good job. This may lead to great non-monetary benefits for the PBC contractors.

## Case study of horizontal projects

A few case studies on the use of the PBC approach are explained as follows.

### *Canada*

The 1988 PBC contract was a pioneer in outsourcing the transportation asset management work in British Columbia, Canada (ADB, 2018). Before the work was outsourced to a private contractor, the roadway and bridge asset management had been maintained by the government in-house workforce method. The scope of that contract included routine maintenance, winter services, patrolling, and emergency assistance; however, the periodic maintenance, such as pavement rehabilitation and resurfacing activities, was excluded. After the first PBC contract was successful, this type of contract was replicated in other provinces, such as Alberta and Ontario, in 1995 and 1996, respectively (ADB, 2018). British Columbia and Alberta maintained all of their highways using PBCs in 2006 and Ontario maintained more than half of their highways using PBCs (ADB, 2018). The provinces started with three-to-five year PBC contract durations and then progressively increased up to 10 years (ADB, 2018). The reported cost savings obtained through shifting toward the PBC method ranged from 10% to 20% (Pakkala, 2005; ADB, 2018). As British Columbia moved from an in-house workforce method to PBCs, 10% cost savings (estimated) were achieved (Pakkala, 2005; NCHRP, 2009).

### *United States of America*

In the United States, the first PBC transportation asset management contract was awarded to VMS Inc. in Virginia in 1996 (ADB, 2018). The contract duration of that project was 5.5 years (ADB, 2018). Under that contract, per ADB (2018), the PBC contractor was responsible for maintaining a number of assets on a 1,250 mile (over 2,000 lane km) highway. The purpose of the contract was to improve the LOS of the roadway assets with lowered cost. Virginia's DOT estimated cost saving was 16% over the contract period of 5.5 years (ADB, 2018). The assets included pavement surface, roadway drainage system, bridge assets, vegetation and aesthetics, traffic services, emergency response services, and snow and ice control. The PBC contractor came up with new materials to improve service life as well as invented a mobile road 'patcher' (ADB, 2018). The Virginia DOT was satisfied with the contract. To improve the quality of the services, the PBC contractor also trained a number of subcontractors who provided services to the state. After the first PBC project was successful in Virginia, other state DOTs also replicated the PBC approach in the United States.

## *Malaysia*

In Malaysia, the PBC approach was introduced by the Public Works Department of the Ministry of Works in early 2000 (ADB, 2018). A large road network was divided into three separate contracts. They were 1,912 miles (3,075 km), 2,588 miles (4,000 km), and 4,663 miles (7,498 km) long (ADB, 2018). Each of these works was a 15-year contract term, but the parties had price negotiation options every five years (ADB, 2018). Each of these contracts was an area-wide contract, which included periodic and routine maintenance activities of all the road networks in three areas except toll roads. The main objectives of introducing the PBC approach were to secure long-term funding for road maintenance works and to maintain the level of service in a better condition. At the end of the contract, the contractors' work performance was well received by the Public Works Department, and the target objectives were met (ADB, 2018). In 2014, the Public Works Department initiated the process to extend the PBC contracts for another 10 years (ADB, 2018).

## *A case study of a research project*

There are limited studies conducted in the area of PBCs as compared to other areas. In this section, we discuss the important findings of a study titled, "Framework of Performance-based Contracting for Chip Seal and Striping Maintenance Activities". Chip seal is a preventive roadway maintenance activity that is commonly used in US states and is one of the highest expenditure activities for most state DOTs. To date, state DOTs

**Figure 3:** Chip seal on US 93 Clark County, Nevada.

*Source*: Shrestha *et al.* (2020).

are facing problems with using the PBC approach for chip seals because there is a lack of studies on this area (Shrestha, 2016). Figure 3 shows the chip seal work on US 93 in Clark County, Nevada performed by a private contractor under the traditional contracting method (Shrestha *et al.*, 2020). The second picture in Figure 3 shows how the crack width was measured during a site visit of chip seal work in the Nevada project (Shrestha *et al.*, 2020).

One of the important findings of the study was development of a framework to execute PBC for chip seals. The framework presents critical issues for implementing chip seals using the PBC approach. This framework also guides state DOT engineers and managers in identifying critical factors for each of the issues. The identified issues and critical factors may help states in preparing technical specifications for PBC chip seal works (Shrestha *et al.*, 2020).

The framework was divided into three life-cycle project phases: contract document preparation phase, contract procurement phase, and contract implementation phase. In each phase, to execute the PBC method, the key issues and their critical factors are presented in the framework (Figure 4). For example, one of the key issues in the contract document preparation phase is how much in incentives and disincentives should be provided for the chip seal contractors. The findings of the study were in a range of 4–5%. One of the reasons for presenting a range of incentives and disincentives, rather than a fixed percentage, was because the amount of

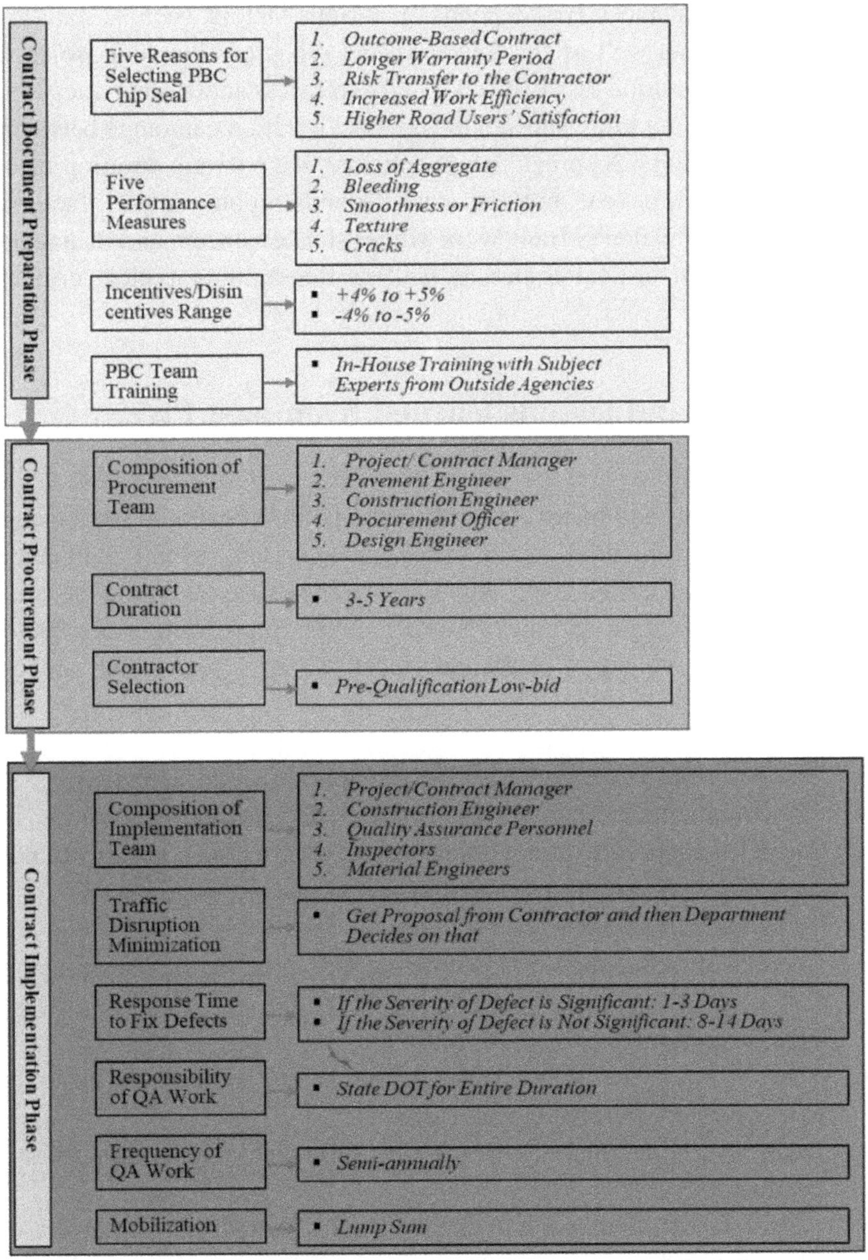


**Figure 4:** A framework to implement PBC chip seal contracts.

*Source*: Shrestha (2016).

incentives and disincentives depends on various factors, such as state-to-state decisions as well as the locations and types of chip seal projects. Similarly, the duration of a PBC chip seal contract is another big question. The findings of the study showed the contract duration can range between three and five years. As previously mentioned, the contract duration for a typical PBC chip seal project also depends upon various factors. Therefore, the developed framework guides state engineers and managers on 13 identified critical issues as well as determining typical critical factors for these issues.

## Challenges and lessons learned from past PBC projects

Several US states and many countries globally have benefited from using the PBC contract method. While using this approach, several challenges have also been identified, especially when agencies are executing the contracts for the first time. Studies (NCHRP, 2009; Stankevich *et al.*, 2009) found the following key challenges faced by agencies, which are as follows:

- The government does not positively support this new approach.
- The contractors are not familiar with the PBC method.
- Agencies had a lack of experience with PBC, or the contract did not meet objectives when it was implemented for the first time.
- The parties involved needed training.
- There were challenges in estimating the cost savings with using PBCs.
- PBCs required secure long-term funding.
- There was a lack of capacity for contractors to complete an accurate asset inventory before starting PBCs.
- Agencies should make sure there is sufficient competition among contractors.
- Agencies should understand that the parties should work proactively.
- There was a downsizing of the staff in an agency.

- A partnership must be established between two parties.
- Performance indicators of an activity must be identified.
- An appropriate incentive/disincentive mechanism must be established.

The lessons learned from previous projects are important to make future projects successful. Some lessons learned that were collected from previous PBC studies (NCHRP, 2009; Shrestha, 2016; Zietlow, 2008; Zietlow, 2004; Shrestha and Powers, 2018) are presented below. They were collected from contracting agencies in North America.

- Make sure the contract language (including performance standards) is very clear to understand.
- Conduct regular meetings and have strong communication between parties to solve issues if any arise.
- Have a partnership between the agency and contractor.
- Make the contractor fully responsible for what they do.
- It is important to have both rewards and penalties for the contractor.
- Conduct random inspections (from qualified individual) with no grace periods.
- In the contract procurement phase, hold a pre-bid meeting, and set the contract duration for as long as the agencies are comfortable.
- Define performance measures clearly, and an independent third party should measure them.
- Provide training to the contractor.
- Pay higher amounts to the contract in the starting months and then linearly after that.
- Secure long-term funding.
- Implement pilot projects to test feasibility and increase parties' experience (take a gradual approach).
- The contract duration should be long enough to maximize potential benefits.
- Understand the cost savings may not be achieved immediately.
- Collect the experience of existing PBC contracts of the same activity.

## Chapter summary

Performance-based contracting (PBC) is a newer outsourcing method, which uses performance specifications. PBC focuses on the end result of a contractor's work, and most contracting agencies provide incentives (rewards) and disincentives (penalty) based on how the PBC contractor performs the work. The amounts of reward and penalty to the contractor depend upon the contracting agency and work type. The key factors that drive contracting agencies to shift from traditional method-based contracting (MBC) to the PBC approach are as follows: opportunity to save costs; easy to achieve project objectives; PBC contractor's projects last longer; less administrative cost with contracting agencies; opportunity to foster innovation. A number of case studies of prior PBC projects show that shifting to this approach saves project costs between 10% and 30%, depending upon a number of factors, including the type of project and contract duration.

Globally, over 39 countries (developed and developing countries) have implemented the PBC method, and many countries plan to implement this method to maintain their roadway assets. Although many countries took advantage of using the PBC method, contracting agencies also faced challenges during the implementation of this method. The key challenges were as follows: governments are not fully supportive of this newer approach; contractors are not familiar with the performance specifications; securing long-term funding can be a challenge.

From past PBC projects, agencies also learned a number of lessons. The important lessons were as follows: making sure the language in the contract documents is not complicated; developing a good partnership between the contracting agency and contractor; providing rewards for good work and penalizing bad work.

# About the Author

**Kishor Shrestha** is an Assistant Professor of the School of Design and Construction at Washington State University (WSU). Kishor earned his bachelor degree in Civil Engineering from Tribhuvan University before earning his master's in Construction Management and his Ph.D. in Civil Engineering with a focus on Construction Engineering/Management from the University of Nevada, Las Vegas. He teaches Estimating I, Estimating II, and Temporary Structures at WSU. His research focuses on the domain of Innovative Contracting Method — Performance-Based Contracting (PBC), Life-Cycle Cost Analysis (LCCA), and Benchmarking of Construction/Maintenance Methods. Kishor partners with states and industries solving their problems in his research areas. He is the author of 1 book, 8 peer-reviewed journals, and 16 conference papers. Kishor is also the reviewer of the Journal of Legal Affairs and Dispute Resolution in Engineering and Construction, Journal of Automation in Construction, Journal of Construction Innovation, Journal of Civil Engineering and Architecture, Associated Schools of Construction, and Construction Research Congress. Kishor is also a Professional Engineer (Construction) and LEED certified individual. Prior to coming to WSU in the fall of 2019, Kishor was in the Civil Engineering department at the Montana Technological University for three years. Kishor also served as an instructor for Construction Engineering Mechanics (Statics) at the University of Nevada, Las Vegas during his graduate study in 2016. Kishor has around nine years of construction industry experience in Nepal and in the UAE before starting his graduate study in the US. During that period, he was involved in Heavy Civil and Building Construction projects as an engineer representing contractor and as consultant as well.

# Chapter 12

# Contractual Aspects of Integrated Project Delivery

Islam H. El-adaway* and Rayan H. Assaad†

**Missouri University of Science and Technology*
*†New Jersey Institute of Technology*

**Abstract**

This chapter focuses on integrated project delivery (IPD) within the construction industry through presenting and studying various contractual related concepts including integrated delivery, partnering, and relational contracting. To this end, this chapter defines IPD and discusses the reasons contributing to its evolvement and popularity. Moreover, this chapter compares IPD to other traditional contracting methods as well as highlights its benefits to the construction process and its stakeholders. This chapter also presents the barriers and challenges associated with the implementation of IPD. Furthermore, this chapter examines how the integrated delivery is incorporated in nationally and internationally standard forms of construction contracts. Finally, this chapter offers insightful discussions and concluding remarks on IPD.

## Introduction

The global construction sector contributes between 5% and 9% of the gross domestic product (GDP) (Assaad and El-adaway, 2020b). The construction sector plays a significant role in driving the activities of multiple sectors including manufacturing, transportation, mining, and agriculture (Donkor, 2011). In fact, construction is considered as a key contributor to the development of economies and industries (Assaad and El-adaway, 2020a). Nevertheless, construction is a complicated sector that is subject to continuously changing conditions (SFAA, 2018). The construction sector suffers from a high number of contract disputes, which contributes to an inability to turn material inventory to cash and thus leading to liquidity (Tserng *et al.*, 2011). One of the key causes leading to conflicts, claims, and disputes in the construction industry could be attributed to the fact that projects are subject to numerous risks and uncertainties (Khalef *et al.*, 2020; Othman, 2008). The construction industry has been plagued with many issues for a long time, which are still not adequately addressed (Assaad *et al.*, 2020d). One of such challenges is the fragmentation of the industry, which results due to the need for close interaction and collaboration between different entities such as owners, architects, design professionals, consultants, builders, subcontractors, suppliers, vendors, manufacturers, and other project players (El-adaway *et al.*, 2017; Assaad *et al.*, 2020b). In fact, each project participant in the construction sector cares mainly about their own interest(s) with little (if any) consideration to the interests of other project parties or to the overall project (Ellison and Miller, 1995; Cheng and Li, 2004). This fragmented nature of the construction sector contributes to numerous conflicts, claims, and disputes between the parties. In fact, the main causes of construction disputes include the project parties' failure to comprehend or conform with their contractual responsibilities and the project participants' failure to adequately manage the contract (Arcadis, 2019).

Given today's complex construction businesses (Assaad and El-adaway, 2021c) and dynamic contracting environment, many efforts have been directed to rely on relational contract agreements to establish a best-for-project culture to avoid the centralization of decisions, which is often detriment to the project and to a philosophy that offers a good construction atmosphere (El-adaway, 2013). That said, partnering and integrated project delivery (IPD) foster a relational contracting environment. To this end, partnering is considered as an effective alternative method to traditional or classical delivery approaches to address the different challenges faced by the construction industry.

It is worth mentioning that the concept of partnering has been present since the 80s as the US Army Corps of Engineers implemented it in their projects to avoid or minimize the antagonistic relationships present between the project participants (Harmon, 2003), which were created by the traditional contracts and delivery methods. Similarly, IPD started in Europe and was embraced by the United States later on as an approach to support project alliances.

There is not a standard or unified definition for IPD that is widely accepted and used by the entire sector (Kent and Becerik-Gerber, 2010). The differences between existing definitions are seen in relation to their approaches, team formation, and contract arrangements (Sive, 2009). Nevertheless, there exist many common features between the multiple versions or types of IPD (Ahmad *et al.*, 2019). That said, this chapter provides the definitions that are more commonly used in the industry and that include the agreed upon commonalities. According to Pheng (1999) and Neo (1998), (1) the partnering process aims to create working relationships between the project stakeholders based on a mutually established philosophy of communication and commitment and (2) partnering establishes a culture such that claims and disputes are prevented by teamwork and trust, a cooperative relationship is promoted for the benefits of all team members, and a successful completion of the project is achieved. In simple words, partnering is the reliance on a teamwork philosophy for resolving the different issues and problems that take place in construction facilities and projects (El-adaway, 2010). Moreover, according to Cheung *et al.* (2006), the following two propositions are at the substantive core of the relational contract theory: (1) a contract is basically a cooperative or collaborative social behavior and (2) agreements with substantial relational aspects are the predominant type of contracting.

The American Institute of Architects (AIA) (2007) provides the following definition for IPD: "a project delivery approach that integrates people, systems, business structures, and practices into a process that collaboratively harnesses the talents and insights of all project participants to optimize project results, increase value to the owner, reduce waste, and maximize efficiency through all phases of design, fabrication, and construction." Kent and Becerik-Gerber (2010) stated that the following is needed to have an efficient IPD approach: (1) early involvement of all key project participants; (2) multiparty contract or agreement; (3) having shared risks and rewards. Figure 1 shows a simple representation of the contractual relationships between the key projects parties for the common project delivery methods in the construction industry.

Integrated delivery is a subject of great interest in the construction sector (El Asmar *et al.*, 2013). Thus, this chapter introduces the reader to the different aspects of partnering, relational contracting, and IPD. These concepts are considered to fall under one umbrella that is commonly referred to as 'integrated delivery'. This chapter focuses on integrated delivery in the construction sector in specific. This chapter presents the reasons behind the evolvement of integrated delivery and the different elements or aspects that played an important role toward its growth. Afterwards, this chapter offers a comparison between traditional or conventional delivery and integrated delivery and discusses the principles of integrated delivery method. As stated before, the adoption of integrated delivery has considerably increased in recent years due to its significant advantages in different aspects. Thus, this chapter reviews the benefits associated with the use of integrated delivery. However, like any other delivery method, integrated

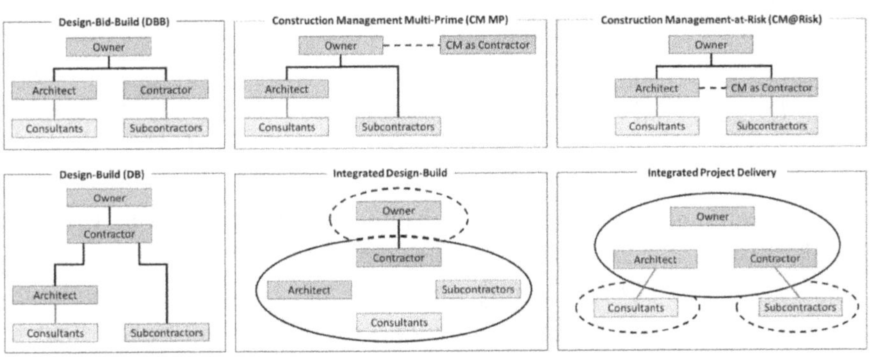


**Figure 1:** Common project delivery methods in the construction industry.

delivery faces multiple challenges. Hence, this chapter also considers the shortcomings or barriers associated with the use of integrated delivery. Subsequently, this chapter examines how integrated delivery is incorporated in nationally and internationally standard forms of the construction contract. Finally, this chapter provides discussions and conclusions related to integrated delivery in the construction industry.

## Reasons behind integrated delivery evolvement

The design–bid–build model was the most commonly used in the United States during the 20th century because public procurement policies encouraged this delivery method since the 40s (Miller *et al.*, 2000). Nevertheless, as construction facilities became more complex, the design–bid–build method led to the creation of multiple cultures within the industry which led to inefficiencies, high costs, and high fragmentation (Kent and Becerik-Gerber, 2010). To address these problems, another delivery method evolved in the 60s which is the construction management model that offered the project owner's value but it did not solve the issues resulting from fragmentation. In the 90s, the design–build delivery method came into existence, and it proved its positive impacts on project performance. During the same period when the design–build model was evolving in the United States, a project delivery method called 'project alliancing' was implemented in Australia on many public projects (Noble, 2007). The project alliancing approach calls for the implementation of a collaborative model to align the goals and incentives of the construction team members. Also, the project alliancing model is popularly called 'IPD' which has currently surfaced in the United States (Kent and Becerik-Gerber, 2010).

With today's complex contracting environments (Assaad and Abdul-Malak, 2020b) as well as the evolving dynamics within the construction industry, traditional contracting methods are considered unsuitable. O'Connor (2009) highlighted that traditional agreements firmly delineate responsibilities in a way focused on the consequences of failure, which strengthens self-defensive behavior and leads to mistrust among the project parties. O'Connor (2009) adds that traditional contracting methods inadequately allocate risks to the parties that are least capable of managing them; thus, traditional agreements presume a high degree of clairvoyance when allocating risks.

Although many efforts have been directed to enhance coordination in the construction industry, traditional contracts provide major obstacles toward having effective and efficient collaboration and communication. These obstacles are especially experienced in large-scale projects that are becoming common in today's construction, are subject to multiple uncertainties, and are highly complicated (Assaad *et al.*, 2020a). In other words, despite the constant positive changes and amendments to traditional contracts, these agreements still reflect a tools-based process that only calls for project completion using segmented and unconnected delivery (El-adaway, 2013). According to El-adaway *et al.* (2017), the traditional project delivery methods might be preferred only when (1) the duration of the project is reasonably short, (2) high interaction, communication, coordination, and collaboration are not required in the project, and (3) little chance or opportunity exists for future collaboration and cooperation between the project parties. Nevertheless, traditional project delivery methods cannot be used for partnering since they promote fragmentation and hierarchical contracting practices which are unsuitable for an integrated delivery of construction projects (Grajek *et al.*, 2000).

Mosey (2001) presented the following statements in regard to using the traditional contracting methods: (1) *contractors* believe that projects' owners care mostly about getting their projects for the best price as clients do not possess the knowledge about the project's details and particularities, and they do not always follow or implement the expert decision; (2) *clients* assume that their projects usually experience cost and schedule overruns due to the contractor's lack of proper planning and poor performance; (3) *suppliers* think that they need to be engaged in the final phases of the construction process and that they are constantly required to reduce their quotations, charges, or prices. While allocation of contractual obligations under traditional contracts is known to the project parties (Assaad *et al.*, 2021a), such traditional relationships result in many inefficiencies including conflicts and disputes between the project participants. O'Connor (2009) highlighted that the reliance on traditional contracts and classical project delivery model usually lead to inferior outcomes. For instance, many capital projects are not completed within budget, and schedule overruns are constantly experienced in construction facilities and projects (Assaad and Abdul-Malak, 2020a).

That said, multiple professional construction institutes and organizations have published reports and studies to enhance the traditional contractual relationships and project delivery methods by focusing on aspects such as risk sharing and allocation, integrated decision-making, design coordination, change order valuation and management, early involvement, and dispute resolution, among others. In summary, with the increasing trend toward the incorporation of all participants into a single team, the IPD model obtained growing acceptance and interest (Kumaraswamy *et al.*, 2005).

## Comparison between traditional delivery and integrated delivery

This section provides a comparison summary between traditional and relational agreements as summarized by El-adaway *et al.* (2017).

### *Parties to the contract*

Under traditional agreement, the contract is considered a two-party agreement between the contractor and project's owner. The contract is a multiparty agreement between the following project participants including the owner, architects/consultants/design professionals, builders, vendors and specialists. On the other hand, under relational or integrated agreement, the contract is a multiparty agreement between the following project participants including the owner, architects/consultants/design professionals, builders, vendors, and specialists.

### *Parties' involvement*

Under traditional agreement, minimal involvement of the project's owner is needed. All parties to the contract, except the consultant, are engaged very late after the completion of the detailed design phase and preparation of the bidding and tender documents. On the other hand, under relational or integrated agreement, the involvement of the project's owner is encouraged at all the stages and the owner shall have a qualified representative who is actively involved in the decision-making process. All parties to the contract are involved in the early phases of the project (such as conceptualization or planning stage) and provide active guidance and input.

## *Identification of project cost*

Under traditional agreement, the cost of the project is determined through a competitive bidding or tendering approach that allows the owner to obtain the lowest cost for the project (Assaad *et al.*, 2021b). Nevertheless, this could result in the selection of claim-oriented contractors who reduce their bid values and rely on claims during the construction phase to obtain more than the agreed upon contract price. On the other hand, under relational or integrated agreement, the cost estimates of the project are updated during each one of the design stages. Nevertheless, the target cost or budget of the project is determined by the agreement between the entire team, and the cost might or might not be specified as a guaranteed maximum price for the project's owner.

## *Value engineering*

Under traditional agreement, these contracts or agreements do not necessarily include a value engineering clause or stipulations for the parties to share the cost savings. Also, value engineering is considered to be limited due to the contractor's late involvement. In other words, the contract is mainly able to save by changing the way activities are executed on the job site rather than by providing constructability feedback during the early design phases. On the other hand, under relational or integrated agreement, value engineering is highly recommended to be implemented by all team members since any cost savings are shared between the parties. In addition, the early involvement of the project parties provides them with the capabilities to innovate in multiple design aspects and work procedures for the best interests of the project.

## *Conflict of interest*

Under traditional agreement, the main goal of the project's owner is to minimize the costs and time at completion. On the other hand, the contractor's key objective is to earn the greatest possible profit. Thus, this results in a conflict of interest since each project participant is concerned with their own goal(s) regardless of the project's interests or outcomes. Also, traditional contracts focus on the consequences of failure and individual responsibilities which result in adverse relationships between the project participants. On the other hand, under relational or integrated agreement, this type of agreements focuses on aligning the interests and objectives of all project parties while achieving project success. This is attained through incentives and profit sharing in addition to the definition of preset objectives for the entire project such that cost savings, enhanced quality, and better project outcomes result in incentive payments for all parties. This minimizes the occurrence of conflict of interest and focuses on directing all project participants to work together toward successfully executing the project. Thus, this establishes a win-win situation for all project parties without focusing on the consequences of failure.

## *Communication and sharing of information*

Under traditional agreement, communication is mainly characterized as hierarchical and there is selectivity in information sharing since the project participants are concerned that other project parties might use such information against them later. Also, project parties might intentionally withhold information to utilize it later in their claims. On the other hand, under relational or integrated agreement, communication exists horizontally between the team members and is supported at all levels. In addition, sharing of information that benefits the project is promoted because withholding information can result in loss of profit for all project parties. Moreover, the use of BIM and technology is promoted for information sharing and coordination of activities to minimize inefficiencies, misunderstandings, and discrepancies.

## *Risk management*

Under traditional agreement, the contractor is the main party that is responsible for risk identification, assessment, and management by developing its own mitigation and control practices or strategies as well as defining the contingency amounts for each risk. This often results in an inflated contract amount. Moreover, each project participant avoids taking responsibility for the identified or unforeseen risks by contractually transferring them to the other project participants. On the other hand, under relational or integrated agreement, all project participants are responsible for risk identification, assessment, and management. This leads to active input from all team members as well as to innovation in dealing with, mitigating, and controlling the risks. In addition, separate contingency amounts might be allocated apart from the contract price in case of risk occurrence. In relation to that, risks are allocated to the project participants who are the most competent to handle with them rather than to the parties that are in a weaker position to deal with them.

## *Claims and disputes*

Under traditional agreement, claims and disputes are considered inevitable and constitute one of the ways through which the builder can recover some monies related to the executed work and make a profit. In fact, the construction sector is subject to numerous conflicts, claims, and disputes that contribute to considerable schedule and cost overruns (Elsayegh *et al.*, 2020). In addition, the contract provisions include the agreed upon dispute resolution mechanisms to minimize the impacts of claims and disputes and to allow the parties to address these impacts easily and efficiently (i.e., through a resolution hierarchy, using mediation, and using arbitration that is less time-consuming than litigation). Nevertheless, the claim-oriented nature of traditional agreements results in having adversarial relationships that can negatively affect the outcomes of the project.

On the other hand, under relational or integrated agreement, claims and disputes are avoided through the incorporation of contract stipulations that waive the parties' right to claim against one another except in very specific circumstances where claims cannot be avoided. In addition, the project teams (being the core group, the project management team, or the project executive team) act as an extra layer for early dispute resolution between the different participants. These groups include members from each organization and thus they are likely to reach a compromise that satisfies all concerns. The fact that many rights to claim are waived helps in promoting a culture with more trust and allows for effective and efficient information sharing and collaboration.

## *Performance evaluation*

Under traditional agreement, given that the contractor finished the activities with acceptable quality, his or her performance is mainly assessed based on time and cost. On the other hand, under relational or integrated agreement, all team members are evaluated based on predetermined KPIs and goals that are specifically set to assess performance in relation to project progress and success. This approach offers enhanced reflection for every member's performance. The performance of team members is often tied to incentives.

## *Organizational culture*

Under traditional agreement, traditional contracts fit organizational cultures where roles and obligations are clearly defined, and each party focuses on its preset obligations without focusing on the bigger picture and how the individual responsibilities serve the benefits of the project. Also, project parties try to avoid any additional obligations. On the other hand, under relational or integrated agreement, relational contracts are best for organizations with task-based cultures where the focus is mainly directed to the task at hand (being the attainment of a goal to gain an incentive or meeting a certain target for a KPI or success of the project as a whole). This means that organizations are more flexible to form teams, delegate authority, and take on obligations to attain the objectives of the project.

## Principles of integrated delivery

The principles or success factors of a partnering relationship are trust, transparency, no blame culture, team work, early involvement of integrated delivery teams, constant enhancements as related to performance considerations, and long-term relationships (El-adaway, 2010; Lazar, 2000; Cheng and Li, 2002; Anvuur and Kumaraswamy, 2007). One of the critical principles of integrated project delivery is proper and fair risk sharing among the project parties since risk allocation alone does not address the common and undesired protective and defensive attitudes embraced by project participants in construction projects (Crane *et al.*, 1999).

Hanna (2016) stated that IPD's principles could be divided into two major categories, that is, contractual and behavioral which are described as follows:

- The contractual aspect includes treating the main project parties equally and as a single team, sharing of rewards and risks based on project's results and outcomes, waiver of liability among the main project stakeholders, having transparency between the main project parties, having early involvement of the main parties, relying on intensified design, jointly developing target criteria for the project, and exercising a collaborative and cooperative decision-making process.
- The behavioral aspect includes having mutual trust and respect between the team members, being willing and ready to cooperate, and embracing a culture based on open communications.

It is worth mentioning that the contractual aspects constitute a formal agreement through a written form, while the behavioral elements are considered to be preference-based aspects (Ahmad *et al.*, 2019).

In summary, AIA (2007) stressed that the following nine principles of IPD shall be embraced by all project parties to achieve the benefits of IPD:

- *Mutual respect and trust*: The key project parties (owner, designer, consultants, constructor, subcontractors, and suppliers) shall understand the value of collaboration on an integrated project and be committed to work as a team for the best interests of the project.
- *Mutual benefit and reward*: The IPD model provides many benefits for all team members. In other words, compensations on IPD projects are structured in a way that recognized and rewards the early involvement of the key project parties. More specifically, the compensation is determined based on the value added by the team member(s) and it rewards the behaviors that are focused and directed to the benefits of the overall project like attaining the goals and objectives of the project. In relation to that, innovative business models shall be used on IPD projects to endorse efficiency and collaboration.
- *Collaborative innovation and decision-making*: Innovation on integrated projects is embraced and leveraged when all team members are freely allowed and encouraged to exchange ideas and thoughts. The IPD model assesses ideas based on their merits rather than on the members' role, position, or status. Also, the important and critical decisions are assessed by the entire team and made unanimously to the best possible extent.
- *Early involvement of key participants*: On IPD projects, the main project parties shall be involved from the earliest practical moment. This early involvement can improve the decision-making process through the influx of expertise and knowledge of all key parties. In other words, the overall combined experience and knowledge of the key project participants provides the best outcomes during the early phases of the project where informed decisions possess the highest possible impact(s).
- *Early goal definition*: On integrated projects, the goals shall be developed, defined, and agreed upon in the early phases of the project, and

all project participants must respect them. Also, the IPD model values insights from all team members by adopting a culture that drives and leverages innovation and superior performance and that holds project's outcomes at the heat within a framework of individual participant objectives and values.

- *Intensified planning*: The IPD model provides that increased efforts during the planning phase of the project would lead to better efficiency and savings during the construction phase. The IPD does not aim to reduce the design efforts, but rather to substantially enhance the design results and streamline and shorten the expensive construction efforts and aspects.
- *Open communication*: The focus on IPD projects is on open, honest, and direct communication between all project parties. That said, the responsibilities, roles, and obligations are defined in a clear way and by avoiding a 'blame culture'. This would result in better identification and resolution of issues with no determination of liability. All disputes and conflicts are recognized the moment they happen and are resolved very quickly and efficiently.
- *Appropriate technology*: The IPD model often requires the use of cutting-edge technologies. In fact, technologies are usually specified during the project initiation phase to optimize generality, functionality, and interoperability. Moreover, having open and interoperable data exchanges based on disciplined and transparent data structures are very critical to support the IPD framework. That said, the project participants shall rely on technology which is in compliance with open standards that enhance the communications and collaborations among all project stakeholders.
- *Organization and leadership*: The formed project team shall be viewed as an organization in its own right where all members are dedicated to achieve and attain the goal, objectives, and values set for the project. In addition, the leadership shall be claimed by the most capable team member as related to the project's specific work and services. Moreover, roles shall be clearly determined while removing any barriers that chill open collaboration, communication, cooperation, and risk taking.

It is worth mentioning that IPD principles could be used for multiple delivery systems (AIA, 2007). For instance, projects with construction management at risk or design–build delivery methods do not rely on a multiparty contract but they could be referred to as IPD-ish or near-IPD when they include some of the previously mentioned IPD principles for improved communication, cooperation, and collaboration (Hanna, 2016).

## Benefits of integrated delivery

One of the benefits of integrated project delivery is that it could lead to lower tender prices since contractors tend to submit a low contingency amount because risks are shared between the project parties. This could also lower the chance of having claims-oriented contractors (El-adaway *et al.*, 2017). Moreover, partnering and IPD methods create an atmosphere that is based on trust and that supports communication, cooperation, and collaboration between the parties as well as maximizes the success of the construction mechanism through the organization of teams and the convergence of the participants' interests through the integration of the experience and knowledge of all members in a cooperative and collaborative setting (Zaghloul and Hartman, 2003; El-adaway, 2013; AIA California Council, 2007).

Additional benefits of the IPD method is that it could most effectively facilitate the use of building information modeling (BIM) on construction projects (Kent and Becerik-Gerber 2010). BIM is one of the most prominent technological advancements (Assaad *et al.*, 2020c), and the National Institute of Building Sciences (2015) defines BIM as follows: (1) a digital representation of the physical and functional characteristics of a facility; (2) a business process for generating and leveraging building data to design, construct, and operate the building during its lifecycle; (3) the organization and control of the business process by utilizing the information in the digital prototype to affect the sharing of information over the entire life cycle of an asset.

Kent and Becerik-Gerber (2010) highlighted that the construction phase is where the benefits of the integrated model are mostly realized and they include the following: less change orders, more enjoyable projects, cost savings, shorter schedule, fewer request for information, improved quality, decreased stress and friction, and better productivity. Manata *et al.* (2018) stated that the use of IPD enhances the performance of the team, improves project outcomes, eliminates challenges that can hinder information sharing, improves team trust, and offers an enhanced environment for open interactions.

In summary, AGC (2009) summarized the benefits of the IPD model as related to the three key project participants: owners, architect, and contractors. The benefits to owners include having a faster delivery for the project, having both the designers and contractors together 'own' the owner's program and vision, having continuous design validation and optimization, having zero non-discretionary change orders, having lower final cost, being able to have more building for the dollar, and having the owner advocates everywhere. The benefits for architects include having true alignment with the builder, having joint 'ownership' of the design, having the design quality being the builder's financial goal, eliminating out of pocket losses, being able to reach 85% liability risk reduction, and having significant upside fee potential, among others. The benefits for contractors include eliminating conflicts, being able to participate in the creative process, having higher quality subcontractors, experiencing lower risks, building relationships and repeating clients, unemployment of attorneys, and having better execution (i.e., smarter, leaner, safer, and faster), among others.

## Challenges faced in integrated projects

Despite the many improvements that have been performed in the construction sector during the past decades (Assaad and El-adaway 2021a) and despite the numerous benefits of the IPD method, many barriers, shortcomings, or challenges exist. AIA (2012) published a report on multiple IPD case studies, which includes shortcomings and lessons learned. This report concluded that the IPD method still needs to be improved and many weaknesses need to be addressed since it is considered a relatively new approach. According to the AIA (2012) report, the following challenges were highlighted by the different case studies:

- Absence of clear and definite description and alignment of goals at the early phases of the project can lead to devastating outcomes.
- Absence of adequate planning between the team members at the early phases of the project.
- Absence of trust and continuity, particularly for the owner.
- Non-availability of the owner and contractor during the construction phase of the project.
- Absence of standards for reviewing project's snapshots when using BIM.
- Having multiple architects handling the entire project instead of a single architect.
- Lack of synergy between the contractor and the designer.
- Having unclear contingency money which is not precisely defined (i.e., team's inclination to consider events as changes instead of relying on the contingency pool).
- Having no clear definition of how the contingency savings should be managed and controlled and whether they are part of the incentive pool or not.
- Lack of conducting a scoping phase during the concept design where project participants are reimbursed for the incurred expenses until a mutual agreement is reached about the project's time and budget.
- Not having a separation between costs and profits.
- Combining incentive payment with hourly rates.
- Absence of owner's involvement from the early phases of the project until the construction phase.
- Having unqualified BIM superintendents.
- Subcontractor's foremen not attending the group scheduling meetings.
- Reliance on a hurry-up schedule that can lead to poor team engagement and team-building activities that encourage collaboration.
- Uncertainties related to the iterative designs required for the IPD method.
- Using different software packages for BIM.
- Absence of a facilitator that is responsible for holding meetings, aligning the different interests of the team members, and helping the team work together.

- Not having a boot camp for team formation to prepare for the project and to improve collaboration.

Other shortcomings were also highlighted by different studies. For instance, a study conducted by Bilbo *et al.* (2015) highlighted that some challenges faced on IPD projects include the following: having a high number of individuals who are involved in the decision-making process, which could slow down the process and make it unnecessarily complex; having excessing and extensive meetings, which could lead to meeting exhaustion; having high or unrealistic expectations from IPD participants due to IPD's publicity in the industry; having self-interest exceeding the interest in the project, which is attributed to human nature; having superficial issues related to the subcontractors on the project, which could waste the time of the entire team on something that is not important; having inexperienced subcontractors in relation to open-book dealings, which could lead to inefficient and ineffective cooperation.

In addition, Saal *et al.* (2013) stated that one of the challenges faced is the *we already do it* syndrome which refers to the fact that when many industry practitioners are asked about IPD, they respond "we have always done it" but when asked to explain the process, they reply in a myriad of ways. Moreover, a study conducted by Kent and Becerik-Gerber (2010) stressed that some of the barriers behind not using the IPD method include the following: it does not fit their business model, it is risky to implement as a new system, fear of or reluctance to change, dearth of awareness on the IPD model, business risk, absence of legal structure that supports IPD implementation, lack of insurance coverage, greed, mistrust, and cultural barriers. Moreover, many concerns related to IPD practices were raised by Kent and Becerik-Gerber (2010) in relation to insurance and risk allocation, BIM usage not being advanced to an extent that adequately endorses IPD aspects, the owner not being ready for the IPD approach due to a lack of understanding of such an approach, the parties being unwilling to share their savings, and implementation of processes that are not adequate for the IPD model.

El-adaway (2010) identified a set of other challenges that could be experienced in partnering projects including the following: (1) delays in team building and formation, which could lead to not exploiting the benefits of having early involvement of different project stakeholders such as vendors and contractors; (2) lack of commitment from any one of the members of the partnering team, which could lead to a setting of suspicion and uncertainty as ideas are not communicated and shared with all members and issues are not raised in front of the team members; (3) failure to attain a documented mutual understanding between the members, which could lead to inaccurate interpretations, unmet deadlines, and wasted resources; (4) lack of clarity and openness as related to risks, costs, profits, and joint risk management, which could lead to undermining the entire purpose of the partnering model; (5) failure to achieve an alignment between the partnering agreements and the individual contractual relationships, which could result in the parties reverting to their contracts when issues happen to protect their own and individual interest.

The challenges of the IPD approach do not solely depend on the contract used but also on the organization's business model, the team selection process, and company's and team's culture, and the firm's legal structure, just to name a few.

## Incorporation of integrated delivery in national and international standard forms of the construction contract

The IPD method is incorporated into national and international standard forms of the construction contract. For the international standard forms, the focus will be on the United Kingdom due to its construction industry being highly adversarial and litigious (Pryke, 2006).

### *Relational and IPD contracts in the United States*

The most commonly used standard forms for the IPD method in the United States include those published by the AIA and ConsensusDocs. AIA issues contract agreement for three levels of integrated project delivery: transitional forms, multi-party agreement, and single purpose entity (SPE). AIA (2020) summarizes the difference between these three levels by stating, "*Transitional Forms* are modeled after existing construction manager agreements and offer a comfortable first step into integrated project delivery. The *Multi-Party Agreement* is a single agreement that the parties can use to design and construct a project utilizing integrated project delivery. The *Single Purpose Entity (SPE)* creates a limited liability company for the purpose of planning, designing and constructing the project. The SPE allows for complete sharing of risk and reward in a fully integrated collaborative process."

Under the transitional documents relationships, AIA has the A195 document (*Standard Form of Agreement Between Owner and Contractor for Integrated Project Delivery*), B195 document (*Standard Form of Agreement Between Owner and Architect for Integrated Project Delivery*), and A295 document (*General Conditions of the Contract for Integrated Project Delivery*). The A195 mainly stipulates the business terms and conditions that are specific to the agreement between the owner and the contractor rather than the unique scope of work that the contractor is responsible for; the duties and obligations of the contractor, owner, and architect are specified in the A295 contract which A195 incorporates by reference. The same applies for the B195 which mainly stipulates the business terms that are specific to the agreement between the owner and the architect rather than the unique scope of services that the architect is responsible for; also, the duties and obligations of the contractor, owner, and architect are specified in the A295 contract which B195 incorporates by reference. Finally, the A295 does not only stipulate the duties and responsibilities of the owner, architect, and contractor but also establishes how these project participants will work together during each phase of the project: conceptualization, criteria design, detailed design, implementation documents, construction, and closeout (AIA 2020). It is worth mentioning that A295 necessitates that the project participants use BIM on the project. Figure 2 shows a summary of the different AIA documents under the transitional level.

Under the multi-party agreement relationships, AIA has the C191 document (*Standard Form Multi-Party Agreement for Integrated Project Delivery*) which is used as a single agreement by the owner, architect, contractor, and other main project parties for the design, construction, and commissioning of the project. AIA (2020) provides that this document offers a "framework for a collaborative environment in which the parties

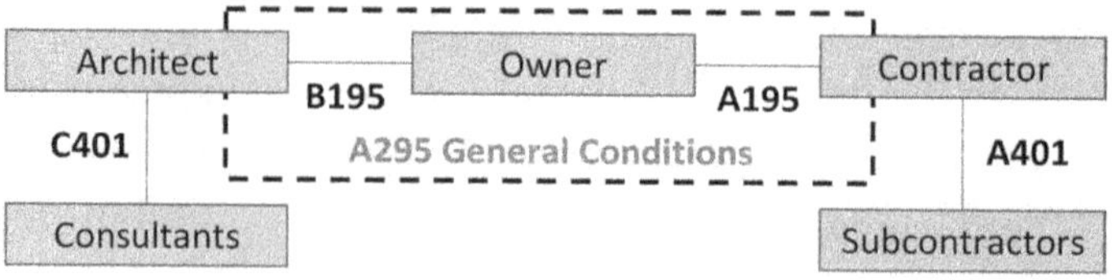


**Figure 2:** AIA transitional documents relationships.

*Source*: Adapted from AIA (2020).

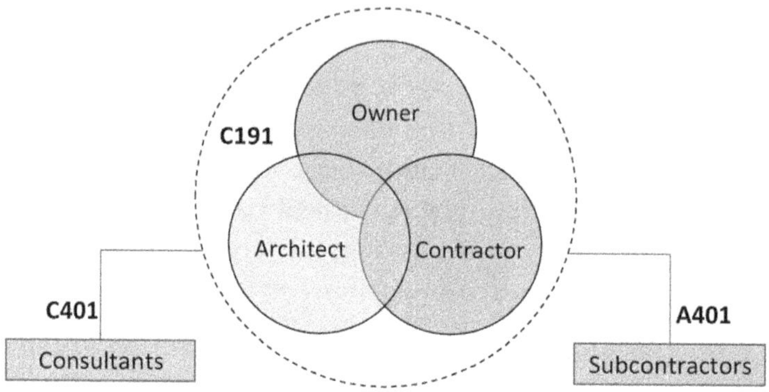


**Figure 3:** AIA documents for the multi-party agreement.

*Source*: Adapted from AIA (2020).

operate in furtherance of cost and performance goals that the parties jointly establish. The non-owner parties are compensated on a cost-of-the-work basis. The compensation model is also goal-oriented and provides incentives for collaboration in design and construction of the project. Primary management of the project is the responsibility of the Project Management Team, comprised of one representative from each of the parties. The Project Executive Team, also comprised of one representative from each of the parties, provides a second level of project oversight and issue resolution. The conflict resolution process is intended to foster quick and effective resolution of problems as they arise." Figure 3 shows a summary of the different AIA documents under the multi-party agreement relationships.

Under the single purpose entity contract relationship, AIA has the C195 document (*Standard Form Single Purpose Entity Agreement for Integrated Project Delivery*), the C196 document (*Standard Form of*

*Agreement Between Single Purpose Entity and Owner for Integrated Project Delivery*), the C197 document (*Standard Form of Agreement Between Single Purpose Entity and Non-Owner Member for Integrated Project Delivery*), the C198 document (*Standard Form of Agreement Between Single Purpose Entity and Consultant for Integrated Project Delivery*), and the C199 document (*Standard Form of Agreement Between Single Purpose Entity and Contractor for Integrated Project Delivery*). The AIA C195 document is an agreement such that the owner, architect, construction manager, and other main project parties become the member of a limited liability company. The main goal of this company is to design and construct a project that relies on the IPD principles. Nevertheless, according to AIA (2020), "To obtain project funding, the company enters into a separate agreement with the owner [i.e., C196]. To design and construct the project, the company enters into separate agreements with the architect, construction manager, other non-owner members [i.e., C197], and with non-member consultants [i.e., C198] and contractors [i.e., C199]. The compensation model in the non-owner member agreements is goal-oriented and provides incentives for collaboration in design and construction of the project, and for the quick and effective resolution of problems as they arise." Figure 4 shows a summary of the different AIA documents under the single purpose entity level.

On the other hand, the ConsensusDocs has two main documents for the IPD method: ConsensusDocs 300 (*Multi-Party Integrated Project Delivery Agreement*) and ConsensusDocs 396 (*Joining Agreement for Integrated Project Delivery*). The ConsensusDocs 396 document is intended to be

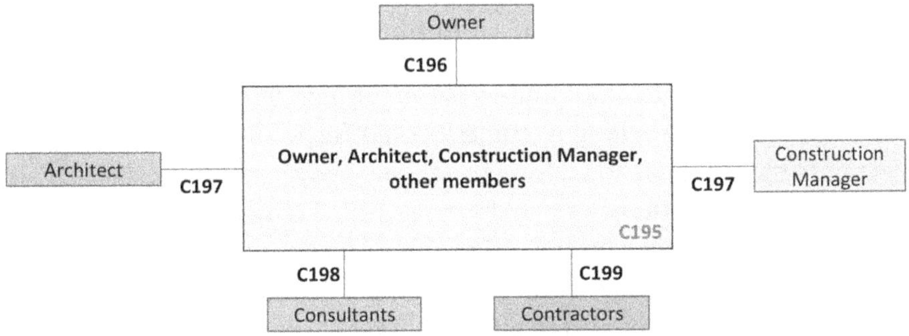


**Figure 4:** AIA documents for the single purpose entity level.

*Source*: Adapted from AIA (2020).

used with the ConsensusDocs 300 document such that all joining parties agree to cooperate and collaborate to boost productivity and produce desirable project performance and outcomes. Under the ConsensusDocs 300 document, the three key project participants (owner, design professional, and constructor) sign the same exact agreement, thus rendering it an integrated form of agreement (IFOA). According to ConsensusDocs (2020), the ConsensusDocs 300 document includes lean principles such as pull planning, pull scheduling, A-3 thinking, target cost, estimated maximum cost, target value design (TVD), TVD clusters, target value clusters, and a risk pool plan (to share risks and rewards). In addition, each project party executes a joining agreement (such as the ConsensusDocs 396). Moreover, the parties are classified into two main categories under ConsensusDocs: (1) risk pool and (2) joining members. As the name suggests, the risk pool members possess a higher risk and reward level compared to the joining members. As far as the general conditions of the ConsensusDocs 300 are concerned, ConsensusDocs (2020) specify that these conditions "include the responsibilities allocated to each party and address critical construction law contract issues. The contract issues include: the parties' relationship and responsibilities to each other, what the agreement is to be, IPD method and Lean construction principles, collaborative project management, validation phase, preconstruction phase, Design Professionals, Constructor's compensation, costs, risk allocation and pool, price estimates, contract time, changes, payment, termination, dispute resolution and mitigation, and much more. Parties can modify the terms of the agreement for their individual project needs and state law requirements."

## *Relational partnering contracts in the United Kingdom*

Due to the adversarial and litigious nature of the construction industry in the United Kingdom, the government asked Sir Michael Latham to provide possible solutions to address the challenges faced by the industry. In response to that, Latham prepared a report titled 'Constructing the Team' in 1994 which included around 30 recommendations focusing on designing a partnering approach for the entire supply chain. These recommendations provided the ground for the formation of multiple UK partnering standard documents such as the Project Partnering Contract (PPC) of the

Association of Consultant Architects (ACA) and the New Engineering Contract's (NEC) Option X12. Davison and Ryan (2008) highlighted the following:

- The ACA's PPC 2000 is a multiparty agreement between the project's contractor and owner (that is, two-party) which stipulates that the owner's consultants and the project's subcontractors are considered parties of the agreement. In addition, the PPC 2000 contract document calls for collaboration and includes procedures which it innovated to stimulate its users to complete their projects in an efficient and effective way by being part of a Partnering Team. The PPC 2000 documents establish a Core Group that is responsible for the different processes and mechanisms related to the planning, procurement, and delivery of the construction activities. Moreover, the PPC 2000 contract documents provide an uncommon direct legal relationship between the different entities of the supply chain. Furthermore, the PPC 2000 document is designed to allow for changes and to motivate all project stakeholders to propose changes that can help in the project execution and delivery or that can enhance the operation and maintenance of the completed facility. Also, the PPC 2000 contract document stipulates for the establishment of key performance indicators (KPIs) that are related to monetary incentives and shared savings. That said, the PPC 2000 document provides that payments could be processed based on the KPIs, and it stipulates an agreed maximum price (AMP) such that neither overheads nor profits can be obtained, to make sure that collaboration will not be used as an excuse for a suboptimal execution of obligations or for a cost plus payment framework.
- The NEC partnering option (Option X12) is considered as an additional aspect to the primary or core options that are stipulated for the two-party agreements. More specifically, the Option X12 aims to improve the partnering and collaboration tools that are incorporated by the core NEC by changing agreements with collaborative aspects into partnering contracts. Using the NEC's Option X12 means collaborative construction is at the heart of the relationship between the project participants by the addition of many processes. That said, the

project participants can rely on NEC's Option X12 to promote a better relationship with the subcontracted designers or equipment suppliers necessitating design development.

## Discussions

Alternative project delivery methods, such as IPD, were introduced to address some of the issues experienced in traditional procurement delivery, mainly design–bid–build (Ahmad *et al.*, 2019). Though the IPD model has some challenges and shortcomings, it has great potential reflected by its benefits to the project, teams, and the overall industry. In other words, the benefits of IPD outnumber its challenges, and IPD showed superior performance when compared to other project delivery methods. More specifically, the study conducted by El Asmar *et al.* (2013) collected data and quantitative information to assess the performance of IPD projects compared to projects executed using the following delivery methods: design–bid–build, design–build, and construction-management at-risk. This previous study showed that the IPD method provides statistically significant improvement in 14 metrics across the following six performance areas: quality, schedule, change orders, environmental performance, communication among project participants, and cost performance. Ibrahim *et al.* (2020) also concluded that the IPD method outperformed the design–bid–build delivery method in 11 metrics and that the IPD method outperformed the construction-management-at-risk method and the design–build delivery method in two metrics each. Further, most studies have showed some evidence that the more collaborative project delivery methods (such as IPD) are superior to, and better than, less collaborative systems (such as traditional delivery methods) (El Asmar *et al.*, 2013). Due to the numerous advantages of IPD, it is expected that the reliance on delivery method will continue to increase in the future (Assaad and El-adaway, 2021b).

Saunders and Mosey (2005) and Mosey (2001) stressed that the following issues could still arise even when a proper contractual framework is implemented: (1) the project's owner not being involved in an active and constant way with the members of the relational contracting team; (2) team members being resistant to change; (3) not involving specialists, manufacturers, and vendors in the early phases of the project; (4) not adhering to or establishing the partnering schedule; (5) having unclear incentives, KPIs, and goal compensations; (6) following a dominant individualistic approach for the management and control of risks; (7) having a partnering advisor that is not engaged in an active and constant way with the project-related decisions and activities.

Despite that, each party to the IPD agreement has its personal insurance for liabilities and the entire contract or team would not inevitably be covered (El-adaway *et al.*, 2017). To address the challenges faced by IPD projects in relation to insurance, El-adaway (2013) recommended the use of risk retention groups to offer the needed coverage against third-party claims.

El-adaway *et al.* (2017) highlighted that a successful partnering and IPD framework shall tackle the following key aspects during each one of the project's three main phases: (1) *preconstruction and procurement stage*: the project participants must adequately define the objectives and deliverables and shall develop clear processes related to the documentation and commencement of the project; (2) *construction stage*: the project participants must establish proper practices accessing the job sites, design development, payments, changes, record keeping, collaboration, schedule updates, communication, and conflict resolution; (3) *postconstruction phase or maintenance and operation*: the project participants must ensure satisfactory handing of the facility's main intent/purpose, shall be involved in the punch list operations, should settle the due monetary amounts, and must solve any claims or disputes between the different parties.

Each one of the members of the relational team is responsible for ensuring that at least one of its employees attend and participate in the meeting of the core group to fulfill their agreed upon obligations (Association of Consultant Architects, 2008). It is worth mentioning that for the core group to take decisions, all members attending the meeting shall provide their agreement. According to Mosey (2001) and Saunders and Mosey (2005), the core team must work (1) as a brake on (i.e., does not allow) unilateral decisions, (2) as the recipient of concerns from any member of the relational group, (3) as an impartial assessor of the project's KPIs, (4) as a review board that provides impartial assessments in case of conflicts, and (5) as one level in the problem solving and dispute resolution hierarchy. Furthermore, the core group is very important to maximize the opportunities of partnering and IPD. Having a partnering advisor will further enhance the partnering efforts and initiatives.

To embrace and leverage an efficient and effective partnering and IPD environment, a culture change and long-term strategies are needed. Training could play an important role in achieving those changes by equipping associated industry practitioners with the principles of partnering and the IPD model. Such training shall promote a unified understanding of partnering and IPD strategies to establish a proactive involvement in minimizing bureaucracy and determining innovative procurement approaches.

## Conclusions

Traditional agreements or contracts do not solve the roots of the issues faced by the construction industry, but rather they address the symptoms only. The reason behind this is that construction contracts are traditionally formed based on discrete or individual transactions without factoring in the agreements needed to sustain the relationships between the project parties, especially in complex projects. That said, collaborative methods have evolved to face the problems resulting from the use of traditional agreements. Partnering introduces different types of changes including cultural, managerial, and operational by integrating owners, architects/designers, contractors, suppliers, and other project participants.

A good IPD contractual framework for multi-party partnering shall not only have the key project parties (client, architect or design professional, and builder) together under a single contract but shall also include other participants including project/construction managers, consultants, vendors, and subcontractors. That said, all project parties shall together constitute the members of a single relational contracting team where members work together as well as work individually with trust, fairness, and mutual collaboration for the overall project's benefit and according to their stipulated roles, experience, and obligations. In addition, the members of the relational group shall form a core team that must meet periodically to review and drive the project's activities and progress.

A partnering contract must ideally include the following key aspects: (1) mutual coordination and cooperation, teamwork, fairness, and shared financial motivation; (2) explicitly defined obligations, duties, responsibilities, and roles through an integrated document; (3) agreement on risk allocation and pricing project's changes in advance; (4) possessing some sort of flexibility in relation to payments with defined payment entitlements and incentives; (5) having adequate processes for avoiding or reducing conflicts and claims or following a speedy dispute resolution. To leverage partnering and IPD, the project participants must recognize the long-term benefits from such relationships as to build trust between the team members and to enforce the relational contract to the best extent possible.

Existing standard IPD contract documents address different aspects including how decisions are taken, how and when to establish the project's cost, how incentives and compensation are determined, and how changes and contingencies are managed. That said, these standard IPD documents establish a framework that is based on risk allocation (such as insurance, limitation of liability, and indemnity), dispute resolution processes, transparency, and access to documentation and project's records. As far as risk management is concerned, a well-structured partnering contract shall (1) determine risks and identify their potential costs, (2) avoid or minimize risks and the associated costs, (3) insure that risks are affordable and appropriate, and (4) share or apportion the different risks based on the team members that are best capable of managing them.

# About the Authors

**Islam H. El-adaway** is the Hurst-McCarthy Professor within the Department of Civil, Architectural and Environmental Engineering at Missouri University of Science and Technology (formerly known as the University of Missouri Rolla). He is the Founding Director of the Missouri Consortium for Construction Innovation (MO-CCI). Dr. El-adaway's funded work in the area of construction and infrastructure management resulted to date in 155 peer-reviewed journals and conference papers with his graduate students. He has received more than 30 awards and honors at the international, national, regional, and institutional levels for his research, teaching, and service activities.

**Rayan H. Assaad** received his Ph.D. from Missouri University of Science and Technology (formerly known as the *University of Missouri Rolla*). Dr. Assaad is currently an Assistant Professor of Construction Engineering and Management and Critical Infrastructure within the Department of Civil and Environmental Engineering at New Jersey Institute of Technology. Dr. Assaad worked as a Project Manager, and his research has resulted in the publication of over 25 refereed journals and conference papers so far. Dr. Assaad's main research interests include modeling, simulation, and computational methods; modern management techniques and innovative technologies; workforce- and project-related aspects; and infrastructure asset management.

# Chapter 13

# International Contracting

Ying Gao*, Zhuo Feng[†] and Shuibo Zhang*

**Tianjin University*
[†]*Dalian University of Technology*

## Abstract

International contracting has been playing a more and more important role in the global construction market according to Engineering News-Record (ENR). This chapter first gives a big picture of the current status of international contracting practices in global construction markets, which specifically cover the market status, development of delivery approaches, governance, and model contracts. A focused literature review on this topic, based on six selected top journals in the field of construction and engineering management is then conducted. Six themes emerge in international contracting, namely risk management, entry decision, performance, joint venture, governance, and social responsibility. Some key aspects of these themes are elaborated and comments on the key research findings are offered. It is concluded that research on international contracting is still at a budding stage and far from mature. Based on a brief discussion of the current studies and knowledge gaps, this chapter finally offers a future research agenda with regard to the knowledge boundary, corporate-level knowledge, industry-level knowledge, and institutional environments to build a better international contracting body of knowledge, which will facilitate a healthy and sustainable development of international contracting business.

## Introduction

It is generally acknowledged that civil infrastructure and industrial facilities are not only central to economic and social development but also enhance the quality of life, particularly in developing countries (McDermot *et al.*, 2020). According to Global Infrastructure Hub (2017), the demand for global infrastructure investment by 2040 is expected to reach $94 trillion, along with $3.5 trillion being needed to ensure electricity and water development under the United Nations' Sustainable Development Goals. Meanwhile, with the increase of productivity, construction projects are becoming larger in scale and technically more complex (Lavikka *et al.*, 2018). These projects usually involve a vast amount of resource input, including finances, equipment, human expertise, and materials, making it very hard, if not impossible, for a single country or region to cope with these challenges alone (Tetteh and Chan, 2019). In addition, the comparative advantages of cross-national firms in the construction field make efficiency more likely to be achieved if they work on the same project (Zhang, 2018). Currently, global projects with large scales are ubiquitous in international construction markets, and for major contractors, international contracting (IC) is playing an important role in their business sustainability and even survival.

The term "IC" is frequently used in the literature across several related disciplines, such as economics (as in Yarbrough and Yarbrough, 1994), business (as in Lee, 1988), and law (as in Folsom *et al.*, 2012), in addition to the field of construction management. However, the term is often used implicitly and somewhat differently, without explicit definitions. To avoid misleading the readers, the term in this chapter is defined, based on its setting of the Architecture, Engineering and Construction (AEC) field, as follows: *IC is a business model in that an AEC organization contracts for and performs cross-border construction projects, including design and construction services, the supply of materials and equipment, and increasingly in recent years, financing, and operation and maintenance (OM) services, in return for revenues and profits.* An AEC organization that does such cross-border business is often called an *international contractor*, as the term is used in this chapter.

The above definition brings to light several implications. First, it concerns a body of knowledge on the topic of IC that focuses more on the corporate level rather than on the project level. Second, IC exposes contractors to institutional differences due to its cross-border nature, implying more risks, particularly with regard to legal issues and governmental regulations, cross-cultural communications, different technical standards, payment, foreign currency exchange, and foreign tax issues (Loo *et al.*, 2013). Third, geographical distance requires international contractors to be more concerned with logistics and governance issues than those involved in domestic projects, and thus intra-organizational control becomes a salient concern. Hence, international contractors are required to have more distinct knowledge and master a different skill set to cope with these principal features of IC.

This chapter intends to integrate the current and various fragmented IC knowledge areas into a tentative body of knowledge through a comprehensive review of the current status of IC practices, relevant research topics, and findings. At the end of this chapter, a future research agenda is proposed for the purpose of building a more mature IC body of knowledge.

## Practices in international contracting

### *Current status of the international contracting market*

The IC market has been fluctuating due to the intensifying downside risks of global economic growth recently. However, in terms of revenue, there has been an average annual growth rate of 2.11% for the Top 225/250 International Contractors over the past decade, with 2015 being the peak year, reaching a total revenue volume of $544 billion.[1] Of all the international markets for these top contractors, Asia and Australia have been the

[1] All the figures in this section are based on Engineering News-Record (ENR) statistics.

largest market regions in terms of revenue for nine consecutive years, followed by Europe and the Middle East. Although the Top 225/250 International Contractors have been diversifying their businesses in various market sectors, transportation, petroleum, buildings, and power remain the four dominant sectors, accounting for more than 80% of their international business volumes. Very recently, the international market seemed to be shrinking. As reported by the Top 250 International Contractors, the contracting revenue outside their home countries shrank from $487.29 billion in 2018 to $473.07 billion in 2019, and the reduction accounts for 2.9% of the total revenue in 2018. Furthermore, the domestic project revenue of $1.287 trillion in 2019 was reported by the firms as a group, up 12.1% in comparison to $1.148 trillion in 2018. In addition, the IC market has been seriously impacted by the COVID-19 pandemic. Considering that the supply chain in the global construction industry has been disrupted by the pandemic, it is predicted that the IC market will most likely continue to shrink in the coming years, in line with the worsening of the global economy. However, it is believed that IC will grow steadily in the long run.

## *Development of project delivery approaches in international contracting*

IC is a project-based business in which different delivery approaches are applied. A project delivery approach describes how the project participants are organized in their interactions with each other, with the objective of transforming the owner/employer's goals into a finished facility (ASCE, 2012). Traditionally, project delivery in IC is generally classified into two main approaches, namely design–bid–build (DBB) and design–build (DB)/engineer–procure–construct (EPC) (Yu *et al.*, 2017). With the development of the international construction market, the concept of public–private partnerships (PPP) has emerged as a new approach for project delivery over the past two decades (Cui *et al.*, 2018).

In DBB, the project owner assumes the responsibility for the pre-bid and post-construction work, including financing arrangements, design and tendering service (usually outsourced to independent consultants), and operation after project completion. The general contractor is solely responsible for construction.

DB and EPC are synonyms and are often used interchangeably in academic writings (Zhang *et al.*, 2019). However, in practice, the term "DB" is generally used in the building and civil engineering sectors, while EPC often features in the heavy engineering sector. In DB or EPC, the contractor undertakes not only construction but also the design service and provides goods for the project. Recently, a deviated form of EPC — contractor's financed EPC — has emerged in IC and is worthy of greater attention, particularly in overseas projects contracted by Chinese international contractors. In this deviated form of EPC, contractors provide financing services to project owners, in addition to design and build, in two ways. The first type of financing service is buyer's credit, where the contractor provides assistance and coordination only in obtaining project finance, with the project owners being the real borrowers from financial institutions. The second type of financing service is seller's credit, where the contractors borrow money from financial institutions as debtors and use such loans to build the project. The project owner pays back the money as principal, with financing charges, to the contractor after the project has been completed, as per the EPC contract and/or a separate financing agreement.

PPP has also become prevalent in IC in recent years, mainly due to a shortage of funds by host governments, particularly in developing countries. It is an umbrella term that usually includes design–build–finance–operate (DBFO), design–build–operate (DBO), build–operate–transfer (BOT), build–own–operate–transfer (BOOT), transfer–operate–transfer (TOT), etc. International contractors, acting as project sponsors/investors and usually executing the project in the form of a project company (Special Purpose Vehicle), provide an integrated scope of services, i.e., financing, construction, operation, and maintenance, based on a project agreement (concession agreement) with the host government. Under this approach, the international contractor actually plays multiple roles — not merely as a constructor but more as an investor and operator.

This development of project delivery approaches has gone hand in hand with the expansion of the services provided by international contractors, from construction only to an integration of financing, design, build, and operation. Therefore, international contractors have been playing an increasingly integrated role in ICC, particularly for infrastructure development.

## *Governance in IC and the FIDIC model contracts*

IC is targeted to projects which are temporary and inter-organizational activities requiring close communication and collaborative efforts from multiple participants with different legal and cultural backgrounds (Du *et al.*, 2016). These differences in institutional contexts among project participants lead to obstacles and problems in their communication, which further result in poor coordination, misunderstanding, chaos, and even project failure. Therefore, a *private order*, as a good governance mechanism, must be designed to alleviate the effects caused by such institutional differences. Contracts can play such a governing role (Williamson, 2002). Being legally enforceable, binding, and managerially instrumental, they offer "the rules of play" for all the parties to follow while also acting as a guide for parties to work cooperatively (Williamson, 2002). To achieve this purpose efficiently and effectively, standard or model engineering and construction contracts have been developed by various international professional associations or institutions for application in IC, such as International Federation of Consulting Engineers (FIDIC), Joint Contract Tribunal (JCT), Institution Civil Engineers (ICE), American Institute of Architects (AIA), and Associated General Contractors of American (AGC). Of all the model contracts, FIDIC contracts are the most notable as they are widely used and have achieved widespread global acceptance (Baker *et al.*, 2009).

FIDIC is a non-government organization founded in France in 1913 and currently based in Geneva, Switzerland. It is best known for its released model contracts specifically targeting the application in IC markets. For the past several decades, FIDIC contracts have been mandated or recommended for use in projects financed by multiple development banks, such as the World Bank, the Asian Development Bank, and the African Development Bank. The widespread application of FIDIC contracts in IC has even made these model contracts become known as *the international best practice documents* on international markets (Klee, 2018).

FIDIC prepared and released its first model contract in 1957, titled Conditions of Contract for Works of Civil Engineering Construction (known as the Red Book). Thereafter, FIDIC released the second edition of the Red Book as well as the first edition of the Conditions of Contract for Electrical and Mechanical Works (known as the Yellow Book) in 1967. These were followed by the third edition of the Red Book and the second edition of the Yellow Book in 1977 and the fourth edition of the Red Book and the third edition of the Yellow Book in 1987. In 1995, FIDIC released the first edition of Conditions of Contract for DB and Turnkey (known as the Orange Book). In response to the development of delivery approaches and international market changes and for use in the new millennium, FIDIC published its most influential Rainbow Suite of Model Contracts as a reform of its previous versions of contracts. The new suite consists of four books: Conditions of Contract for Construction (first edition of the New Red Book), Conditions of Contract for Plant, DB (first edition of the New Yellow Book), Conditions of Contract for EPC/Turnkey Projects (the Silver Book), and Short Form of Contract (the Green Book). All these model contracts are quite distinctive from the previous models, both in content and format. Since their release in 1999, these model contracts have been playing a dominant role in IC until the present. In 2008, FIDIC also published its DBO Contract (the Gold Book) for use in projects with a contractor not only being responsible for facility design and construction but also the operation of the finished facility with a period of up to 20 years or more before being handed over to the employer.

In 2017, FIDIC released its long-awaited second edition of the New Red Book, the New Yellow Book, and the Silver Book, to be followed by the release of the second edition of the Green Book in due course. Although the second editions generally follow the structures and principles of the first editions, in order to achieve its alleged "improved clarity and certainty", they have far more detailed contractual provisions (almost doubled in length), more time bars in procedure administration, and more reciprocity of obligations between the parties. However, whether these updated versions of the FIDIC main contracts are more welcome than their predecessors remain to be tested in their application in the IC market.

In addition to its traditional main contracts, FIDIC also published supplementary and complementary model contracts, such as Form of Contract for Dredging and Reclamation Works (Blue Book, 2006, 2016) and Conditions of Contract for Underground Works (Emerald Book, 2019) to be used specifically for some IC sectors and Conditions of Subcontract for Construction (First edition, 2011) and Conditions of Subcontract for Plant, DB (First edition, 2019) to be jointly used with the New Red Book and the New Yellow Book.

Indeed, a simple literature review shows that FIDIC contracts have not only been frequently applied in practice, but they have also attracted attention from researchers. Several guide books on FIDIC contracts have been published by scholars and practitioners in recent years, and research topics on FIDIC contracts appear with increasing frequency in journals or magazines. Thus, FIDIC contracts can be considered as a very important building block of the IC body of knowledge and are worthy of further study.

## Current research topics and findings for international contracting: Literature review

Provoked by this salience in the IC industry, researchers have begun to conduct relevant studies, and thus the IC body of knowledge is gradually taking shape. To identify the current status of the research themes and findings, a literature review in construction management field was performed. The review covered research papers published between 2000 and 2020 in six notable journals in the field of construction management, namely *International Journal of Project Management, Journal of Construction Engineering and Management, Journal of Management in Engineering, Construction Management and Economics, Project Management Journal,* and *Engineering, Construction and Architectural Management*. The terms "international" or "overseas" or "global" were searched in the titles of research papers published in these six journals during the period, and 157 research papers related to IC were identified.

By analyzing the keywords in these papers, six themes emerge in IC, namely risk management, entry decision, performance, joint venture, governance, and social responsibility. Since most of the research on performance and joint venture are included in the other four themes, the themes of entry decision, risk management, governance, and corporate social responsibility, which are considered to be the most relevant to IC, are focused on and elaborated in this section.

## *International contracting market entry decision: Three dimensions*

The IC market brings not only opportunities to contractors but also greater challenges and higher risks than the domestic market due to cross-border institutional differences and longer geographical distances. If a contractor makes a wrong decision about market selection or chooses an inappropriate project, this will adversely impact its IC performance and even threaten its business survival. Therefore, the market entry decision is of vital importance for international contractors and has attracted researchers' attention. Based on an analysis of the existing literature, research on market entry decision-making can be considered as a sequential process of the following: (1) *market selection decision*: International contractors decide which market to enter; (2) *entry mode decision*: International contractors decide how to enter the selected market; (3) *bidding decision*: International contractors decide whether to bid for a project in the selected market and if yes, how. The following sections of our literature review elaborate on these three dimensions.

### *Market selection for international contractors*

As an old Chinese saying goes, "One false move can lose the whole game." For international contractors, choosing the wrong market can have a devastating impact. Therefore, it is important for their strategy to decide which market to enter. This has aroused the research interest of some scholars, as identified below. Their research topics can be mainly categorized into the following three groups:

- Developing theoretical frameworks to assess the competition status of international construction markets. Based on the enterprise niche theory, Zhao *et al.* (2017) used niche width and niche overlap to evaluate the international construction market as well as contractors' competition situation. The results indicated that international contractors' niche widths differ across geography and product dimensions, and four competitive groups emerged. Very recently, Zhao *et al.* (2019) established an evaluation model to investigate the level of market concentration and competitive intensity of the IC business field, which contributes to assessing the competition status in overseas markets for international contractors. Meanwhile, Lee *et al.* (2017) developed a framework to evaluate each target market's current status in terms of the market opportunity, business environment, market experience, and the likelihood of project success. The research findings can assist contractors in making better decisions when addressing overseas market selection problems.
- Identification of factors influencing the market selection decision of international contractors. Specifically, Kim *et al.* (2013) concentrated on the impact of financial valuation on entry or exit time in an IC market, proposing a model of investment valuation. Research findings can guide the international contractors to not only price the waiting value but also exit options so that an investment opportunity being erroneously computed could be avoided. Chen *et al.* (2016a) identified the factors that influence the selection of international market by large Chinese contractors, and they indicated that large Chinese contractors prefer to enter host countries with considerable market potential, small cultural distances, and high levels of country risk and competitive intensity. Based on Indian international construction firms, Viswanathan and Jha (2019) conducted a similar study and presented important factors affecting international market selection, such as geographical and cultural distance between host and home countries, market potential and risk in host country, as well as size and international experience of the firm. The research findings from these two studies suggest that international contractors from different countries may have their own specific preferences.

- Investigation of the impact of critical factors on the market selection decision of international contractors. Chen *et al.* (2016b) empirically examined the relationship between state capitalism and the internationalization behavior of Chinese state-owned construction enterprises and concluded that Chinese state-owned construction enterprises with higher state capitalism level are more likely to enter the markets of developing countries. Meanwhile, based on international construction firms of India, Viswanathan and Jha (2019) identified important influencing factors for selecting international markets and examined the relationship between these factors and the selection decision of international markets. The results showed that the considerable market potential and low risk rating of the host country affect the market selection of Indian construction firms. Very recently, Gao *et al.* (2020) examined the potential effect of cultural distance on the market selection of Chinese international contractors. Their findings indicated that large cultural distance hinders the entry of Chinese international contractors to such a market.

### *Market entry mode selection for international contractors*

International market entry mode is an institutional arrangement to establish and conduct international business (Jia *et al.*, 2017). Selecting the most appropriate entry modes for targeted overseas markets is crucial for international contractors' sustainability in the construction arena around the world (Gundes and Aydogan, 2016). Since 2008, market entry mode has gradually been identified as an important *construct* in international construction research. The specific topics within this can be categorized into the following two groups:

- Identification and classification of potential market entry modes for international contractors. Chen and Messner (2009) identified 10 entry modes for international contractors: joint venture (JV) projects, sole venture (SV) projects, strategic alliances, licensing, JV companies, SV companies, local agents, BOT projects, branch offices, and representative offices. Chen and Messner (2011a) further categorized these basic entry modes in several ways. For example, based on the nature of the relationship between the entrants and other parties, these entry modes can be divided into two types: cooperative modes (e.g., strategic alliances, JV companies) and competitive modes (e.g., SV companies, representative offices). According to the different organizational levels, the modes of market entry can be classified in dimensions of the project level, corporate level, as well as inter-corporate level (Viswanathan and Jha, 2020). By integrating the duration and degree of localization (permanent entry versus mobile entry) with the ownership dimension (wholly owned versus partially owned), Jia *et al.* (2017) established a two-dimensional taxonomy model.

- Identification and classification of factors that contribute to the international contractors' selection regarding market entry mode. Based on the review of the relevant studies, the identified factors affecting market entry mode selection can be classified into the following two levels:
  (1) *Country-related factors*: These factors can be further classified as host-country-related factors and home-country-related factors. Host-country-related factors include cultural distance, target country risk, institutional entry barrier, trade links, trade relations, host market potential, competition intensity, linguistic distance, etc. (Chen, 2008; Jia *et al.*, 2017). Home-country-related factors include home market attractiveness, firm size, long-term orientation, uncertainty avoidance, and multinational experience (Chen and Messner, 2011b).
  (2) *Firm-related factors*: Factors at the firm level have also been highlighted by some scholars and are considered to be more important for international contractors in selecting an appropriate entry mode. Factors identified so far include firm size/firm scale, international experience and ownership structure (Chen and Messner, 2011b; Preece *et al.*, 2016; Jia *et al.*, 2017).

## *Bidding decisions by international contractors*

The bid or no-bid decision requires abundant deliberation by international contractors due to the complexity and uncertainty of international construction projects (Li *et al.*, 2020a). Scholars have already conducted a study on the question of the bid or no-bid decision for international contractors, which can be categorized into the following two groups:

- Identification of potential factors influencing the bidding decision for international contractors. Dikmen and Birgonu (2004) argued that the attractiveness of the project and the competitiveness of a contractor are two important factors that affect bidding decisions and developed a neural network model based on these two factors to rank international projects to prepare project priority lists for international contractors' bidding decisions. Han *et al.* (2005) investigated the influence of international contractors' risk attitude on the bid decision behavior. This research revealed that in choosing between risky opportunities and sure payoffs, participants tended to show weak risk seeking in the profit situation while strong risk seeking in the loss situation, that is, the participants showed the behavior of loss aversion. Low *et al.* (2015) examined the effect of organizational culture on international bidding decisions in the context of Malaysian international contractors. This study revealed that organizational culture indeed affects international bidding decisions, but it is not a dominant factor. Li *et al.* (2020a) identified 41 influence factors for international bidding decision and found that "contractor's capability" and "country risk of the host country" are the most important factors.
- Development of bidding decision model or bidding decision process for international contractors. Wang *et al.* (2019) developed a capability-based risk assessment model to evaluate a firm's capabilities and international construction project risk significance values, facilitating project selection and bidding decisions for international construction firms. Li *et al.* (2020b) built a comprehensive framework to support the bid decision process in international construction projects, consisting of five stages in the following sequence: assessing country risk vulnerability, assessing ability to perform, assessing project risk vulnerability, assessing profitability, and assessing probability of winning. To effectively deal with the uncertainty and make rational bidding decision, Li *et al.* (2021) further proposed a six-stage bidding decision model using the evidential reasoning approach, adding the stage of assessing the general objectives for complex bidding decisions in the international construction setting.

Market selection and entry mode decision are rooted early in research on the manufacturing sector. On the basis of the literature review, it is found that studies on market selection for international contractors have been appearing with increasing frequency, although they are still far less common than expected in the construction sector.

It is also important to note that the notable studies pertaining to international market entry decision-making dedicated to the construction industry have been conducted against the background of emerging economies (e.g., China and India). More studies in the context of other countries or regions are required to make research comparisons and thus facilitate the generalization of the IC body of knowledge.

## *Risk management in international contracting: Proactive and reactive*

International construction projects are more complex due to cross-border institutional and geographical differences, among others, as compared to domestic projects, implying more risks for international contractors. Failures in risk evaluation and mitigation will hinder the success of the project and endanger contractors' business sustainability. Therefore, risk management is perceived to be more crucial for international construction projects. Risk management can be divided into two types: proactive and reactive. Proactive risk management refers to preventive strategies that are taken before a risk event occurs, such as identifying risk factors, assessing risk effects on the project qualitatively and/or quantitatively, and finally, taking some measures to avoid risks. Reactive risk management focuses on emergency measures after an accident so that risk losses can be minimized. For example, the project participants could develop risk mitigation strategies to achieve optimum risk return (Han *et al.*, 2008; Wang *et al.*, 2004; Zhi, 1995).

### *Proactive risk management in international contracting*

Risk management requires proactive measures to defend against and reduce the probability of potential impacts before adverse consequences can occur. Proactive risk management begins with the identification of risks that international contractors may be exposed to. Zhi (1995) divided the risks of overseas construction projects into internal and external aspects. External risk is the uncertainty related to the regional market or environment, while internal risk is the uncertainty of the behavior of stakeholders. In addition, risk could be also classified into three primary areas according to El-adaway *et al.* (2018) and Liu *et al.* (2016), namely project level, market level, and country level. Based on such a classification, Hastak and Shaked (2000) identified 73 international project risk factors from these three levels. Dikmen and Birgonul (2006) recognized 45 risk factors of international projects and classified them in accordance with the project features and country conditions. Besides, Viswanathan and Jha (2020) focused on the Indian experiences of embarking on international construction projects and identified the associated critical risk factors. They found that the critical risk factors are, with the ranking order, project-specific risk factor, the economic and market-specific risk factor, the firm-specific risk factor, and the political-specific risk factor. Steffey and Anantatmula (2011) identified the critical success factors of international projects, namely cultural risks, political risks, virtual risks, and regional risks.

Risk assessment also helps contractors to formulate reasonable goals and choose appropriate risk strategies. Hastak and Shaked (2000) developed an international construction risk assessment model (ICRAM-1), which is helpful for evaluation of the potential risk by analyzing risk at the macro, market, and project levels. Based on the three dimensions of project, society, and nature, Chen *et al.* (2011) proposed a multi-criteria decision model to assess the environmental risk, aiming to help avoiding and eliminating damages and loss resulting from natural disasters in international airport projects. Considering the interrelationships between various risk factors, Dikmen *et al.* (2007) used the fuzzy risk assessment method to qualitatively evaluate the cost overrun risk in the bidding phase of international projects.

Even though various risks exist in international projects, many of them can be prevented if well-prepared management strategies are in place. Loo *et al.* (2013) found that the commonly adopted proactive risk mitigation measures include local partner engagement, local officers' networking, appointment of experienced personnel, as well as guarantees regarding firms' project management capabilities. Wang *et al.* (2004) classified international construction project risks into three categories through a survey approach, namely country-related risks, market-related risks, and project-related risks. After a careful risk evaluation, researchers proposed multiple proactive measures in response to these risks including appropriate feasibility study, insuring political risks, and maintaining a good relationship with the local government. Wang *et al.* (2016) proposed that contractors should strategically and cooperatively manage the risks in project implementation, i.e., partnering with other project participants to improve their organizational capability. Viswanathan *et al.* (2020) offered three measures of risk mitigation, that is, pre-project planning, local participation and contract selection, and used structural equation modeling to analyze the effects of these measures on project performance. The results showed that local participation serves as the most effective measure to affect project success, followed by contract selection and pre-project planning. Fluctuation of the foreign exchange rate is a notable financial risk in international contracting. Kapila and Hendrickson (2001) suggested that effective measures of exchange rate risk mitigation could be borrowing and foreign exchange futures hedging.

### *Reactive risk management in international contracting*

Reactive risk management concentrates on the risk response of international contractors when the risk occurs. Risk response concerns mitigating the actual impact in the shortest possible time. The management of multidimensional risks in international construction projects is crucial to both increasing the profit and securing new projects, which is able to create sustainable growth. For political risk, Deng *et al.* (2014) suggested four political risk response means for international contractors, namely implementing a political risk action plan, making claims for payment and/or extension of time (EOT) under certain clauses, gaining support from the home country, and resorting to international organizations.

In the field of IC, there are few studies on the combination of these two risk management strategies. Using the dynamic meta-network analysis method, Wang *et al.* (2018) built a two-stage framework of risk control management consisting of project objectives, risk events, risk factors, and project stakeholders. In the stage of risk assessment, comprehensively evaluating and sorting the risk factors can be carried out by integrating multiple types of networks. In the stage of risk responding, the analysis of project stakeholders' impact on the meta-network contributes to forming effective risk intervention measures for the respective stakeholders. For example, after an accident, communication and coordination between stakeholders should be strengthened, and timeliness and correctness of information should be ensured.

The research on proactive risk management has been very mature, from risk identification and risk evaluation to risk aversion strategies. In contrast, the attention given to reactive strategies of risk management has been relatively less. As a matter of fact, not all risks can be proactively mitigated, especially in IC markets. Therefore, effective reactive strategies of risk management for international contractors need to be further identified, and how to integrate such proactive and reactive measures so that risk management efficiency can be enhanced is worth studying.

## *Governance in international contracting: Inter- and intra-organizational*

Governance is particularly important for IC because it entails a more complex relations network, that is, both inter-organizational and intra-organizational. Inter-organizational governance deals with the relations between the contractor and its upstream and downstream counterparts (employer/client, subcontractors, sub-suppliers, etc.), while intra-organizational governance deals with the in-house principal–agent chain composed of headquarters and overseas project representatives. For horizontal inter-organizational networks, this can breed opportunism and conflict due to incomplete contracts and diverse interests (Lumineau and Henderson, 2012; Zouher *et al.*, 2014). For the in-house principal–agent chain composed of headquarters and overseas project divisions, the overseas project representatives generally have limited risk management capabilities and available resources, and their goals may be different from those of the headquarters, resulting in some opportunistic behaviors (Han *et al.*, 2019).

Therefore, governance becomes pivotal in IC to mitigate opportunistic risks, promote cooperation, and ensure correct decision-making of the overseas project representatives.

### *Inter-organizational governance in international contracting*

Inter-organizational governance here refers to governance between the contractor and their upstream and downstream counterparts (employer/client, subcontractors, sub-suppliers, etc.). Inter-organizational governance includes both contractual and relational governance (Lumineau and Henderson, 2012). Contractual governance, also known as a "formal contract" or "legal contract", refers to the extent to which inter-organizational relationships are "governed by a formal and written contract which explicitly stipulates the responsibilities and obligations of each party" (Cao and Lumineau, 2015, p. 17). Research on contractual governance in IC focuses on analyzing some standard contract forms for the international procurement of projects, such as FIDIC contracts and NEC contracts. Ndekugri *et al.* (2007) identified the change of the engineer's duty in the FIDIC New Red Book in 1999 and discussed how control can be implemented by the employer with the help of the engineer in an FIDIC contract. Besaiso *et al.* (2018) found that the standard forms of a construction contract (FIDIC/NEC) can serve as a tool to minimize disputes in international projects. Inspired by the functional perspective of the contract and considering the contract provisions as serving the three functions of control, coordination, and adaptation (Schepker *et al.*, 2015), Chen *et al.* (2018) examined FIDIC model contracts in 1999 to explore the effect of project delivery systems (PDSs) on contract design. They found that PDSs affect the complexity of coordination provisions but not that of control provisions, while the impact of PDSs on the complexity of adaptation provisions depends on the sub-class. For example, PDSs affect the complexity of Environmental Changes provisions, but barely influence that of Task Changes provisions.

In addition to the above-mentioned contract governance, relational governance is also indispensable for project participants. Relational governance, also called the "relational mechanism", refers to the extent to which inter-organizational relationships are "governed by social relations and shared norms" (Cao and Lumineau, 2015, p. 17). It is recognized in the current literature that trust is the most common relational governance type (Cao and Lumineau, 2015). In the international construction industry, relationship governance also includes favor exchange, quality communication, and emotional relationship establishment (Lin *et al.*, 2020). Lin *et al.* (2020) showed that relational governance affected contractors' adaptability to foreign situations in undertaking overseas projects, and international environmental complexity could strengthen the impact. Shen *et al.* (2017) revealed that trust was conducive to interface management in international EPC projects both directly and indirectly through enhancing inter-organizational openness and communication. Some scholars have also examined the interaction between contract and trust in governing international construction projects. For instance, Gad *et al.* (2017) showed that neither culture nor risk affects the choice of the dispute resolution method, while trust does. Girmscheid and Brockmann (2010) considered international construction joint ventures and found that formal control mechanisms are more important at the inter-organizational level, while trust is more important at the intra-organizational level.

## *Intra-organizational governance in international contracting*

Intra-organizational governance in IC mainly refers to the headquarters or parent company's management control of overseas project divisions. As mentioned above, due to their salient nature, international projects are laden with greater risks compared to domestic contracting projects. However, some overseas project divisions have limited risk management capabilities and available resources. If the headquarters or parent company does not exert certain control and coordination on these overseas project divisions, the latter may respond to risk improperly or make wrong decisions that may even lead to the failure of the whole project. Moreover, the overseas project division sometimes has different goals from the headquarters, resulting in some behaviors or actions of the overseas project division that may prevent the headquarters from achieving its strategic goals. Therefore, as an essential instrument for headquarters or parent firms, management control in IC is helpful to achieve their goals and promote the performance of overseas projects (Han *et al.*, 2019).

Management control refers to the process of one entity using mechanisms, such as authority, bureaucracy, and culture, to influence the behavior and outcome of another entity (Geringer and Hebert, 1989). Management control consists of three dimensions: management control focus, management control mechanism, and management control intensity (Geringer and Hebert, 1989). Management control focus, also called the management control scope, refers to the scope of activities over which the headquarters exercises control (Geringer and Hebert, 1989). Management control mechanism is defined as those measures or strategies adopted by the headquarters to influence the behavior and results of its subsidiaries (Geringer and Hebert, 1989), while management control intensity describes the extent or degree of control exercised by the headquarters (Geringer and Hebert, 1989). Among the identified research papers, only two relate to management control in IC, and both of them focus on the management control mechanism. Neves and Bugalho (2008) used a case study approach to investigate what mechanisms are adopted by emerging international construction firms to control their subsidiaries/project divisions. They found that the headquarters control foreign activities through a home-based bureaucratic mechanism, but they gradually realized that cultural and social control should be also emphasized to respond to the complexity of international business.

International joint ventures for infrastructure investment projects have recently aroused scholars' research interest. Liu *et al.* (2014) suggested that such joint ventures may complicate the control issues in the consideration of their parent firms (as investors) from different countries, which are different in their strategic objectives and cultural backgrounds. Han *et al.* (2019) focused on how parent firms should exert control on international joint ventures in overseas infrastructure investment projects and found that task interdependence had a positive effect on parents' behavior control and clan control, while the relationships between goal incongruence and behavior, clan, as well as outcome control presented as an inverted U shape.

As indicated above, management control is very important for headquarters in IC; however, insufficient attention has been paid to studying it in IC, except for the two references cited above. Moreover, these two studies only focused on the management control mechanism, with no or little attention to control scope and intensity. In fact, international construction firms need to know "what to control, and how to control". That is, before choosing an effective management control mechanism, international construction firms need to decide which activities of overseas project divisions they need to control and which they do not because overseas project divisions need some autonomy for making quick decisions, and excessive management control cannot deal with complex and changeable external environment quickly, leading to low efficiency and low project performance. Therefore, determining the management control scope and intensity for international construction firms (headquarters) is worth studying in the future.

## *Corporate social responsibility in international contracting*

The significance of corporate social responsibility (CSR) has been increasingly emphasized in international business studies (Wu *et al.*, 2015). Disclosure of CSR reports by international contractors has become a recognized practice, particularly by listed companies. For example, based on content analysis, Lu *et al.* (2016) investigated the CSR reports of the top 50 international construction companies ranked by ENR to systematically analyze the prospects and trends of CSR disclosure. This study found that companies from developed countries maintain a higher level of CSR disclosures than those from developing countries. The insights offered by this study can help international contractors to understand the idiosyncrasies of CSR disclosures and future reporting exercises. Another study evaluated the CSR of international contractors reporting on disaster resilience issues. Using systematic content analysis, Lim *et al.* (2021) developed a framework to evaluate CSR resilience and compared the CSR resilience issues in the CSR reports, focusing on international contractors from five regions (US and Canada, European Union, Middle East, Asia (i.e., Korea and Japan), and China). This study revealed three findings. First, contractors from Asian countries had the highest level of CSR reporting on resilience issues, while contractors from Middle

Eastern countries had the lowest level. Second, the concerns of clients and the government were equipped with the top priority by the five regions, except that Chinese contractors were putting more emphasis on the concerns of client and community. Third, the community's and employees' concerns were more emphasized by Asian and Chinese contractors.

Currently, the CSR-related literature focuses on the status quo of the CSR practices of international contractors, and quite a few nuanced studies remain to be further explored. For example, what factors determine the CSR practices of international contractors? Which factors explain the heterogeneity of CSR practices of international contractors in different regions? Also, what effects do different CSR practices have on their corporate performance?

## Toward a better IC body of knowledge: Knowledge gaps and research agenda

The construction industry is moving toward globalization, despite some twists and turns. IC has been playing a more and more important role in world business. However, research studies in IC seem quite sparse and its corresponding body of knowledge is still at a budding stage, though some knowledge has been accumulated with regard to international market entry, risk management, governance, etc. The boundary of knowledge areas for IC is not even clearly known yet. Considering the nature of IC, these knowledge areas may potentially cover institutional environments, industry-level knowledge, corporate-level knowledge, and for this purpose, we propose the following questions that need to be addressed for future research:

- *Knowledge boundary*: What is the boundary of IC integrated body knowledge? What is its distinctive feature as compared to general international business, or can it constitute a distinctive subsidiary of international business? Can we develop a coherent theoretical framework for mapping out the integrated body of knowledge of IC?
- *Corporate-level knowledge*: What are the competitive advantages of international contractors? What is the relationship of their governance

structures with performance? What are the risk management strategies of international contractors in such a more and more versatile environment? What is their linkage with performance? How can we measure the capability maturity model for international contractors?

- *Industry-level knowledge*: What are the potential sources of funds for the IC industry, and how does the industry attract more investment? What are the implications of the more stringent requirements concerning CSR, green construction, and anti-corruption in IC markets? What is the impact of IT/BIM application in construction on international contractors?
- *Institutional environments*: What are the impacts of institutional environments on the performance of international contractors? How do international contractors tackle the challenges caused by institutional differences? What is the relationship between institutional environment and governance in IC?

When all these basic research questions are answered and relevant industry practices are summarized, a more mature IC body of knowledge will emerge, which will facilitate a healthy and sustainable development of IC business.

## Acknowledgements

We are grateful to NSFC for its financial support (NSFC Grant No. 71901161, No. 72031008 and No. 71702020). Thanks also go to our Ph.D. students Ms. Jingyan Qi and Ms. Qiuhao Xie for their assistance in literature review.

# About the Authors

**Ying Gao** is an Assistant Professor at the College of Management and Economics at Tianjin University, P.R. China. She received her Ph.D. degree from the University of Hong Kong in construction management in 2018. Her research interests include international construction contract management, management control for international construction enterprises, and governance for public–private partnerships (PPP). Her research has been published in high-quality journals such as International Journal of Project Management, Journal of Management in Engineering, and Transportation Research Part A/Part B. Her book *A Study of Contractual Governance of International PPP Projects* was published in 2019.

**Zhuo Feng** is an Associate Professor at the School of Economics and Management at Dalian University of Technology, P.R. China. He received his Ph.D. degree in management science and engineering from Tianjin University in 2016. As a visiting Ph.D. student, he studied at the University of California, Berkeley, CA, USA, from 2013 to 2015. His research interests include the theory of public–private partnership (PPP), international project management, and project supply chain. His research has appeared in high-quality journals such as *International Journal of Project Management, Journal of Management in Engineering*, *Decision Sciences*, and *Transportation Research Part A/Part B/Part E*.

**Shuibo Zhang** is Professor of International Construction Management at Tianjin University, P.R. China. Professor Zhang obtained his Ph.D. from the University of Hong Kong in International Project Management in 2004. He is a Fulbright visiting scholar at the University of California, Berkeley and a senior visiting scholar at Cambridge University. He had long-term involvement with international construction contracts, project investment and financing, and construction management in teaching, research, and practice. Professor Zhang is FRICS and FCIOB, and serves as the arbitrator for several arbitration courts, including CIETAC, SZAC, and TAC.

# Chapter 14

# Construction Disputes

Richard J. Gebken

*Missouri State University*

### Abstract

Over the last century, the construction industry has earned a reputation as a competitive and adversarial marketplace where disparate parties battle one another in zero-sum game scenarios. As a result of this distributive tendency, mistrust, hostility, and other non-cooperative behaviors frequently result in conflicts between project parties. The prevalence of these conflicts and their root causes are well documented in the literature. If not properly addressed, these conflicts can escalate into construction disputes or, worse, litigation. Evidence of the inefficiencies and waste caused by construction disputes is slowly influencing industry practitioners' choices and contracting practices to reduce the impact on the company's bottom line. This chapter will explore the nature of disputes in the construction industry and present a framework for minimizing their impacts through various prevention and resolution strategies. From efforts to create collaborative project environments to contractual language that aims to resolve conflict in a timely manner, the systems presented in this chapter are tools that aim to minimize the impacts of construction disputes.

## Introduction

The construction industry has been a paradoxical leader in both dispute occurrences and dispute resolution systems for decades (Arcadis, 2020;

Cakmak, 2021; Groton, 2005; Keil, 1999; Michel, 1998). While this may or may not be a desirable achievement, the industry has managed to develop and advance many unique approaches to address the potential risks of disputes (Alaloul *et al.*, 2019; Harmon, 2003; Mix, 1997; Peña-Mora *et al.*, 2003; Rubin *et al.*, 1999; Zack, 1997) and implement delivery methods reducing disputes. Additionally, many of these concepts and systems, including reasonable risk allocation, dispute review boards, stepped negotiations, and partnering have been successfully employed in other industries (Peña-Mora *et al.*, 2003; Stipanowich, 1995; Treacy, 1995). However, the justification for implementing these procedures has been based primarily upon contractual requirements, governmental regulations, court orders, limited previous experience, or basic reactionary instinct, and not on measured cost savings.

The reality is that conflict and disputes are not new to the construction industry. Detractors, both internal and external, have deplored the existence and extent of construction disputes for decades. One industry publication, Engineering News-Record, has published several articles and editorials over the past 30 years on the disappointing performance of dispute prevention and resolution in construction projects. These comments include, "Corporate heads are seeing the cost of arbitration and litigation growing (Editorials, 1985, p. 23)." "Over the past decade, we have documented the rise of litigiousness in this industry, lamenting its cancerous effect on virtually all it touches (Editorials, 1994, p. 58)." "When it comes to the construction industry, the main dispute resolution tool remains a lawyer, and every disagreement still looks like a lawsuit (Editorials, 1999, p. 68)." Finally, "During 20 years as a construction lawyer, I've seen many clients faced with the untenable decision of choosing between capitulating to an unfair demand or going through litigation and arbitration and absorbing legal fees that can approach or even exceed the amount in dispute (Rubinstein, 2018, p. 52)."

Numerous other critics agree that citing disputes and litigation as a resource draining endeavor should not be pursued lightly (Pawson, 2003; Rubin *et al.*, 1991; Steen and MacPherson, 2000; Stickel, 1999). One of the few published figures for construction-related ligation costs estimates yearly expenditures near $5 billion USD with that amount conjectured to increase at 10% annually (Michel, 1998). It is no wonder why the construction industry has been stereotyped as an adversarial and combative industry. For an industry intensely focused on data-driven results, more effort should be exerted on project delivery, procurement, and contracting methods' improvement to enhance not only business relationships but also the bottom line (Adrian, 1988).

# Background

## *The relationship between disputes, claims, and conflict*

Believing the only thing that stays constant is "change" is ironically true for the construction industry. However, when examining this idiom through the lens of construction disputes, it could be appropriately rephrased to the only thing that is certain is "conflict". Still, conflict does not have to result in disputes between construction parties. To understand this important distinction, it is necessary to look at the relationship between disputes, claims, and conflict.

While many experts on the subject have established basic guidelines on how to differentiate between these terms, misperceptions still exist. In fact, some authors and industry practitioners use these terms interchangeably when their meanings are quite different (Gebken, 2006; Wang, 2019).

Brown and Marriott define a dispute as, "a class or kind of conflict, which manifests itself in distinct, justiciable issues. It involves disagreement over issues capable of resolution by negotiation, mediation or third party adjudication (Brown and Marriott, 1993, cited by Yates, 2003, p. 1)." However, this definition contains elements that describe both disputes and claims (Gebken, 2006). On the other hand, the definition for dispute proposed by Institution of Civil Engineers Arbitration Procedure is too narrow. They choose to define a dispute as the point in time when, "a claim or assertion made by one party is rejected by the other party and that rejection is not accepted (Kumaraswamy, 1998, p. 3)." Unfortunately, neither the Brown and Marriott (1993) nor the Kumaraswamy (1998) definition adequately separate claims from disputes.

For the purposes of this chapter, Diekmann and Girard squarely describe a dispute as, "any contract question or controversy that must be settled beyond the jobsite management staff (1995, p. 355)." This definition is similar to that adopted by the Construction Industry Institute (1995). It defines a dispute as, "a problem or disagreement between the parties that cannot be resolved by on-site project managers (Construction Industry Institute, 1995, p. 1)." This method of describing a dispute as the result of an unresolved conflict helps establish the progression that occurs prior to its establishment.

If not all conflicts become a dispute, there must also be an interim step or state that lies between these two concepts. This is where the concept of claims comes into focus. Adrian defines a claim as, "A request by a construction contractor for compensation over and above the agreed-upon contract amount for additional work or damages supposedly resulting from events that were not included in the initial contract (Adrian, 1988, p. 2)." Similarly, Richter and Mitchell define a claim as, "A written statement by one party requesting additional time and/or money for acts or omissions by another during the performance of the construction contract (Richter and Mitchell, 1982, p. 475)." Thus, a claim refers specifically to the case where a problem or issue has been documented and written up for review by another party and still has the potential to be resolved at the field level (Gebken, 2006). A dispute, that may or may not be documented in writing, is impossible to be resolved at the field level.

To clarify this matter, Kumaraswamy (1997) developed a useful graphic to help define the relationship between conflict, claims, and disputes. Figure 1 shows how conflict can lead to both disputes and claims. In addition, it shows that claims can in turn lead to disputes when settlement cannot be reached. Thus, we see that conflict, defined as a disagreement of objectives, priorities, or interest between parties (Yates, 2003), is the root cause of both claims and disputes.

While many have highlighted the inevitability of construction conflict (Cheung and Suen, 2002; El-Sayegh *et al.*, 2020; Ho and Liu 2004; Stipanowich, 1996a), disputes, on the other hand, can be prevented or at least minimized through improvements (e.g., sufficient preproject planning, adequate scope definition, partnering, etc.) and proper claims management procedures (Gebken, 2006). However, when disputes do arise, there are a plethora of options and alternatives for resolution methods.

## Sources of construction disputes

The sources of construction disputes have been identified, catalogued, and categorized for decades in the literature (Cakmak, 2021; Construction

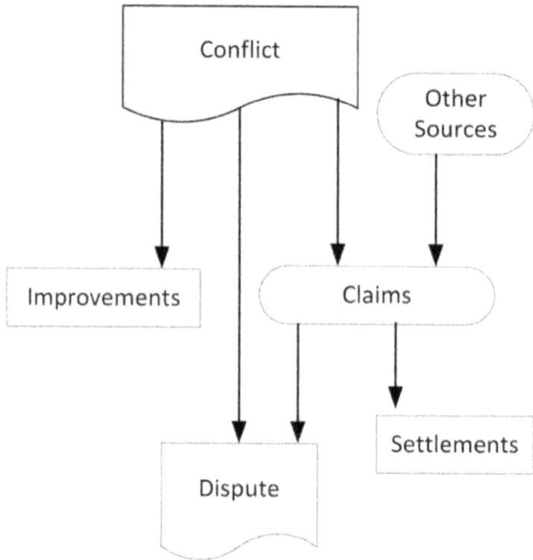


**Figure 1:** Relationship between conflict, claims, and disputes.
*Source*: Adapted from Kumaraswamy, 1997.

Industry Institute, 1995; Diekmann and Abdul-Hadi, 1994; Peña-Mora *et al.*, 2003; Rauzana, 2016). El-Sayegh *et al.* identified 27 possible sources of disputes from the literature and categorized them into five themes, namely, owner-related, designer-related, contractor-related, contractual, and other (2020). Table 1 summarizes these sources of disputes.

Examining each of these categories and related items allows industry practitioners to not only familiarize themselves with the possible causes of disputes but also proactively plan for and anticipate potential problem areas. Like all risk management strategies, identifying and monitoring for potential challenges provides an opportunity to also prevent these issues from ever becoming problematic. Examining the likelihood and severity of construction disputes furthers this risk management approach for construction disputes (Gebken and Gibson, 2006).

## *Construction dispute likelihood*

Research by the Construction Industry Institute (CII) identifying the potential for a construction project to incur disputes and claims was one of the first attempts to proactively manage disputes (Diekmann and

**Table 1:** Source of construction disputes.

| Designer-related | Owner-related | Contractor-related | Contractual | Other |
|---|---|---|---|---|
| Inadequate or incomplete technical plans/ specification | Failure to make interim awards on extensions of time and compensating by the owner | Inadequate site investigation | Contract amendments | Impact of local cultures and social values in the settlement of conflicts |
| Material change and approval during the construction phase | Inadequate early planning of the project | Low financing by the contractor during construction | Contradictory and inaccurate information in the contract documents | Impact on locality in terms of noise, traffic, and pollution/contamination |
| Poor design | Poor financing by the owner | Poor supervision and site management | Differing site conditions | Inappropriate weather conditions |
| Poor preparation and approval of drawings | Slowness of the owner's decision-making process | Poorly defined scope of work | Poorly written contracts | Lack of communication and coordination between parties during construction |
| Time limitation in the design phase | Variations initiated by the owner (additive/deductive) | Shortage and unproductive workers | | Modifying legislation and regulations |
| | | Underestimation and incompetence of contractors | | Obtaining permit/approval from the municipality/ different government authorities |
| | | Unsuitable leadership style of construction/ project manager | | |

*Source*: Adapted from El-Shayedg *et al.* (2020).

Abdul-Hadi, 1994). This work formed the basis for the Dispute Potential Index (DPI) (Construction Industry Institute, 1995, 2017). Embedded in this early study is a system of predicting the relative possibility of a project to be dispute prone. The characteristics used to identify this possibility were based upon three logical underlying categories, namely, people, process, and project (Construction Industry Institute, 1995).

The analysis uncovered that the people factor played the biggest part in dispute potential, while the process and project attributes played important but less influential roles, respectively. In addition, Diekmann and Girard (1995) argued that while people were not necessarily the cause of disputes, they exhibited the greatest influence on project disputes performance, more than any other type of project variable. While there seems to be little analytical literature supporting the softer side of construction disputes (people-related issues), their impacts are uniquely important in field operations. The ability of field personnel to resolve disputes at the lowest possible level allows the project operations to continue with minimal disruptions (Gebken, 2006). Perhaps, the absence of disputes where quality people are on-site and in the project office is a reason why their effects are not easily quantifiable. Again, this would support why most of literature refers to the effects of increased scope, differing site conditions, and inadequate bid information as the main causes of construction disputes.

The process-related factors of construction dispute causes appear much more frequently in the literature. In fact, much attention has been given to construction contracts as both a cause and a possible solution for avoiding construction disputes. Construction contracts have been the major focus of academic journals (Cho, 2020; Jergeas and Hartman, 1994; Komurlu and Arditi, 2017; Semple *et al.*, 1994), industry white papers (Birr and Scnchez, 2017; Frano, 1996), textbooks (Hughes *et al.*, 2015; Mastin *et al.*, 2019; Russell and Jaselskis, 1992), and even government initiatives to decrease the amount of disputes, claims, and litigation on projects (Federal Facilities Council and National Research Council, 2007; Office of Government Commerce, 2002).

Surprisingly, the part of the DPI framework where the most quantitative data are available is within the project-related factors area (Construction

Industry Institute, 1995, 2017). While project-related factors were found to be the least influential on construction disputes in the DPI research, their causes are detailed in the literature frequently. Of particular note are the factors of design complexity, construction complexity, and site limitations. In research by both Diekmann and Nelson (1985) and Semple *et al.* (1994), the major source of construction disputes, and hence claims, was a combination of design errors and scope increases. Anywhere from 50 to 72% of the claims in the studies were shown to occur because of this reason, all of which were outside the control of the contractor (Diekmann and Nelson, 1985). Table 2 summarizes the factors identified in each of the three categories as they relate to the likelihood for causing construction disputes.

**Table 2:** Construction industry institute, 2017.

| People factors | Process factors | Project factors |
|---|---|---|
| Owner/Contractor qualities or characteristics | ***Pre-construction planning*** | ***External variables*** |
| Capable management | Input from all groups involved | Environmental issues |
| Effectiveness of responsibility structures | Financial planning | Public interferences |
| Experience with type of project/ success of past projects | Permits and regulations | Availability of capable craftsmen/subcontractors |
| Experience/Competence | Scope definition | Remoteness |
| Motivation (reward structure) | ***Construction contracting*** | Site limitations |
| Interpersonal skills | Realistic obligations | ***Internal variables*** |
| Business relationship between owner and contractor pertaining to team building | Risk identification/ Allocation | Pioneer projects |
| History together | Adequacy of technical plans/specifications | Design complexity |
| Power balance | Formal dispute resolution process | Construction complexity |
| Expectation of future work | Operating procedures | Size |

## *Construction dispute impacts/severity*

While theory and experience agree that disputes waste a lot of money and resources on non-value adding activities, data is limited on what these losses may total. In 2020, the construction industry installed over $1.43 trillion USD of capital projects in the United States alone (US Census Bureau, 2020). Employing more than 7.6 million workers and constituting nearly 7% of the US gross domestic product, the construction industry greatly influences the overall US economy (US Census Bureau, 2020). The unanimous message at an industry forum hosted by the Federal Facilities Council and the National Academy of Construction was, "The costs of disputes are one of the greatest problems facing the industry today (2004)."

Given the amount of capital expenditures funneled through the construction industry each year and the proclivity of the industry toward conflict, it is surprising to note that no current system exists to evaluate and benchmark the performance of dispute prevention and resolution methodologies in the construction industry (Gebken, 2006). However, it is estimated that for each $1 billion USD saved through the elimination of disputes (a mere 20% of the estimated total spent on lawsuits in construction), a total annual gain of 40,000 jobs in construction could be achieved (Michel, 1998).

A first step to minimize the amount of money and resources that are wasted on conflict and not construction is to define a framework to classify the potential cost impacts. In general, most costs can be broken down into three broad categories including direct, indirect, and hidden costs (Gebken, 2006). Within the direct cost category, dispute costs are generally comprised of outside legal fees, expert witnesses/consultants, and filing fees/court costs.

Indirect costs include salary and benefits for staff, management, and in-house counsel. While these costs are generally well understood, they are often underestimated or simply misallocated to other projects, overhead budgets, or change order proposals that may never be approved (Gebken, 2006).

Beyond direct and indirect costs, hidden and/or unquantifiable costs are the most nebulous of impact areas. This category includes lost productivity, rework, damaged business relationships, loss of company focus/momentum, and loss of business volume, to name a few. Examples of the impacts from this category can include the following: employee loss of focus; extensive amounts of time spent digging for errors or mistakes by the other party to defend or offset allegations by other parties; rapidly deteriorating business relationships and/or reputations; increasing executive time spent on resolving disputes; the potential negative effects on employee morale and their personal lives (Gebken, 2006).

Eventually, the business volume of all companies may decline as conflict protracts and evolves into lengthy disputes. On the contractor's side, bonding capacity and cash flows are major issues that can affect how much new work can be started. On the owner's side, new business opportunities and marketplace entry are issues that are all affected as disputes consume valuable time and resources.

## *Background summary*

While some have stated that the quantity and severity of disputes are invariably linked to the overall health of the economy (Arcadis, 2020; Brooker and Lavers, 1997; Flanigan *et al.*, 1997; Yates, 2003), the consensus belief still appears to be that disputes are unwanted occurrences on construction projects no matter what the economic considerations. However, the importance placed upon construction dispute reduction is justified (Loosemore *et al.*, 2000; McGeorge *et al.*, 2007) and a perceived necessary step to improve the industry as a whole.

When disputes are resolved early, direct and indirect costs can be minimal or even non-existent; however, they tend to grow exponentially as the dispute lingers unresolved or moves toward possible litigation. While most of the hidden and/or unquantifiable costs are the most difficult to quantify, they embody some of the most serious risks when disputes do linger unresolved. Preventing and minimizing disputes in the construction industry may not only help further encourage a shift from combative to collaborative project environments but also help raise productivity in a sector that has seen little improvement in the last three decades (Javed *et al.*, 2018).

## Core knowledge, trends, and expectations

### *Moving toward dispute prevention*

Once potential dispute causes are identified, the natural next step to dispute control is to prevent them from occurring. Many of the ideas explored in this chapter are simple in concept but difficult to implement in practice. Items such as accurate record keeping, adequate knowledge of contracts, preservation of contract rights, careful planning and scheduling, and proactive actions are only a few of the ways in which construction disputes can be mitigated from a contractor's point of view (Arcadis, 2020; Cakmak, 2021; Jergeas and Hartman, 1994).

Design and engineering firms can also help to avoid downstream disputes. In Allen's survey on professional practice (1998), the majority of respondents claimed that clear scope definition was the most effective dispute avoidance measure. This was followed by quality work, adequate budget, and adequate time by and for the design team. Additional measures in the design and engineering phases that may also lead to diminished dispute levels include value engineering and constructability studies (McGeorge *et al.*, 2007; Semple *et al.*, 1994).

Other techniques that stretch beyond the design phase of the project are also vital to dispute prevention efforts. These include project risk assessments, a partnering-type project structure, cost allowance for areas of uncertainty, standard contracts to avoid misinterpretations of project risk, teambuilding exercises, and appointing managers and superintendents with good attitudes and strong cooperative skills (Arcadis, 2020; El-Sayegh *et al.*, 2020; Hasanzadeh *et al.*, 2018; McGeorge *et al.*, 2007; Mitropoulos and Howell, 2001).

The value and importance of soft skills should not be overlooked when implementing strategies to help prevent disputes from occurring. Specifically, industry practitioners should concentrate on communication throughout the construction process (Construction Industry Institute, 1995). Whether it is for reviewing the contract, defining scope, managing changes, or resolving a dispute, communication skills — especially early, consistent, and respectful communication — are paramount for dispute prevention and minimization. Project managers are frequently identified as the key project participant where these communication skills should reside. As one construction attorney explained, "There is absolutely no substitute for an experienced project manager who is not afraid to say no, or more importantly, not afraid to say no nicely (Gebken, 2006, p. 245)." Project personnel must be able to take a step back and work on solutions rather than point fingers of blame. The primary goal of these solutions and others is to limit problems and resolve them at their lowest level in the organization.

Aside from communication skills, a second factor for successful dispute prevention is thorough reviews of contract documents early in the project. Suggested execution methodologies include line-by-line reading of the contract between the two parties at a kick-off meeting or a detailed scope review meeting to catch potential items of conflict. Concentrating on developing good project scopes and good subcontractor scopes, depending on perspective, helps alleviate many problems before they can even become construction issues. In addition to the scope review, another key procedure to follow carefully through is basic document management, which is sometimes referred to as the blocking and tackling of the contracts. Correctly identifying the parties, ensuring all the attachments are attached, and verifying everything is executed is fundamental. In theory, if you went to both parties' contract file, the documents would look identical. This ensures that time isn't wasted fighting over the terms of the contract (Gebken, 2006).

Additionally, disputes can be prevented and minimized by prompt notification of changed scope. This can be facilitated by documenting the original anticipated scope at the onset of the project (e.g., escrowed bid documents) and by actively pursuing issues as they arise. Project team members must be prepared to identify potential conflict areas quickly and notify the appropriate parties promptly. Proactively seeking resolutions to conflict is far superior to letting potentially serious problems fester unaddressed (Federal Facilities Council and National Research Council, 2007).

This leads to another best practice for minimizing disputes — integrating change efficiently into the contract. Dealing with change, whether through effective communication, proper documentation, or formal change management processes, is important when trying to reduce and/or prevent disputes from becoming a major impact on projects.

Lastly, avoiding protracted disputes in construction can also be accomplished by including in the contract adequate means to resolve disputes if they do occur. Practices such as alternative dispute resolution, which includes processes like mediation and dispute review boards, can not only serve as a deterrent to dispute escalation but also as a means to eliminate the growth of hostilities between the contract parties.

## *Minimizing and resolving disputes if prevention efforts are insufficient*

For the construction industry, contract theory is an important part of the legal process that has allowed contractual disputes to be resolved in the courtroom for more than two centuries. However, in the last few decades, courtroom congestion and skyrocketing legal costs have opened up many other opportunities for dispute resolution, most of which are outside the courtroom. This change has been referred to as Alternative Dispute Resolution (ADR). ADR is broadly defined as any method by which conflicts and disputes are resolved privately and other than through litigation in the public courts (Kovach, 2004). ADR techniques can include both binding and non-binding procedures.

As such, the development of a virtual sliding scale of alternative dispute resolution techniques has evolved over the years. Some suggestions of the progression of ADR techniques range from negotiation, mediation, conciliation, neutral evaluation, expert determination, adjudication, to arbitration (El-Sayegh *et al.*, 2020; Kellogg, 1999; McGeorge *et al.*, 2007; Office of Government Commerce, 2002). While many of the construction industry standard contracts (American Institute of Architects [AIA], Associated General Contractors of America [AGC], and the Construction Management Association of America [CMAA]) have traditionally focused their alternative dispute resolution verbiage on arbitration, there is a growing movement to utilize less expensive and less combative system. In fact, an industry survey found that 54% of US corporations said, "cost

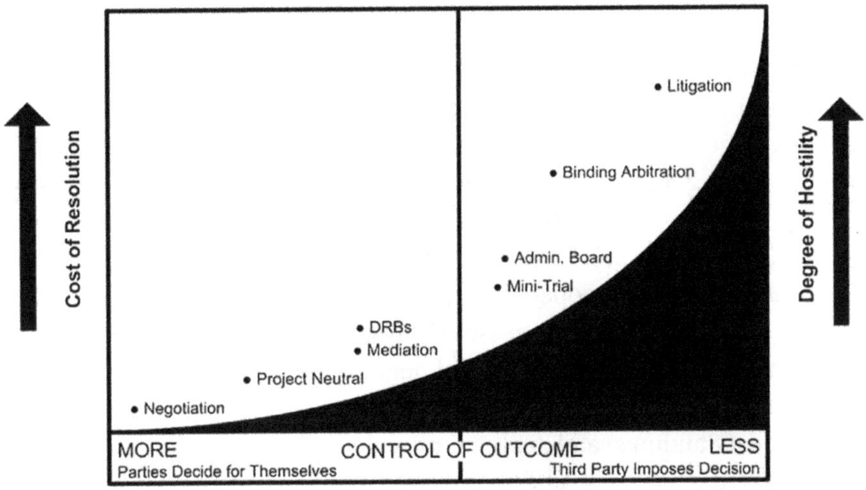


**Figure 2:** Control of outcome vs. cost and hostility of dispute.

*Source*: Adapted from Richter, 2000.

pressures affected their decision to use ADR (Dispute Resolution Journal, 1997, p. 7)."

Figure 2 illustrates a continuum of dispute resolution procedures and their escalating costs and hostilities as compared to the control the parties retain in the process (Richter, 2000).

## Resolving disputes on the jobsite

Dispute resolution techniques that keep control of the dispute in the hands of the parties in disagreement can clearly incur fewer costs during the resolution process and keep hostilities to a minimum. Conversely, disputes that rely entirely on the determination of other individuals (litigation and binding arbitration) are believed to have both higher costs and increased hostilities (El-Sayegh *et al.*, 2020; McGeorge *et al.*, 2007). The following sections will highlight the primary ways in which disputes can be resolved if they do occur on a project. Methods attempting to keep resolution at or on the jobsite will be addressed first. These have the highest potential for minimizing project impacts.

## *Negotiation*

Negotiation has been defined as, "communication for the purpose of persuasion (Goldberg *et al.*, 1999, p. 17)." In the context of the construction industry, negotiation is often the first and last step necessary to resolve disputes. It is generally only when negotiation fails, or fails repeatedly, that a problem becomes a full-blown dispute.

Aside from the prevention of disputes entirely, negotiation is often viewed as a low cost, cooperative endeavor, and favored over other more adversarial and expensive processes (Mays, 2003; McGeorge *et al.*, 2007). Despite the acknowledged benefits of negotiation, it is a process that is not universally successful. One study found that negotiation can fail because of misunderstandings and tactical miscalculations (Loosemore, 1999). Examining the process and communication that takes place during informal negotiations in the construction industry, Loosemore (1999) identifies a trend that hostilities from one party can be retuned with hostilities from the other side, causing a never-ending spiral of increasing conflict and cost. This type of negotiation is often called distributional or competitive.

Another solution is integrative or cooperative negotiating. Oftentimes, the most successful negotiations occur when parties focus on interests and not positions (Fisher *et al.*, 1991). Unfortunately, in the construction industry, positional bargaining tends to be used more often than interest-based bargaining (Peña-Mora *et al.*, 2003).

Most standard contract documents include negotiations as part of the formal dispute resolutions process with the most frequently utilized adoption being the use of step negotiations. Step, or stepped, negotiation calls for escalating levels of authority from both parties the longer a disagreement persists (Groton, 1997). One example is when field level employees who oftentimes have first-hand experience with the issue in dispute raise the negotiation responsibilities to their superiors (e.g., project managers, division heads, or company presidents). In general, contractual step negotiation procedures enumerate both the individuals and the time through which each step of negotiation must follow (McGeorge *et al.*, 2007).

## *Mediation (including conciliation)*

Mediation is a process that is becoming more and more popular in the US construction industry. Mediation can be used to settle disputes on the job site or ones that already involve litigation or other ADR techniques (Mays, 2003; McGeorge *et al.*, 2007). Mediation can be defined as, "a process where a third party neutral, whether one person or more, acts as a facilitator to assist in resolving a dispute between two or more parties (Kovach, 2004, p. 14)." The mediator may or may not be an expert in a given industry, but rather they are presumed to be knowledgeable in the techniques of mediation (Dispute Resolution Board Foundation, 2019; Rubin *et al.*, 1991). Techniques for conducting mediations vary greatly depending on the mediator; however, the mediator does not render a decision but rather helps the parties reach a settlement on their own. In addition, mediation may be ordered by the court during and/or prior to trial. Court-ordered mediation is generally not mandatory; however, parties often follow through on the courts' recommendation to help avoid an impending trial.

Conciliation is a process similar to mediation; yet, a conciliator can propose a solution to the dispute. While offering personal opinions from the mediator is normally frowned upon in other mediations contexts, mediators of construction cases often make evaluations of each party's case during private caucuses. This is done without referring to the procedure as a conciliation meeting. Additionally, construction mediators are frequently expected to propose an offer (normally referred to as a mediator's offer), after the mediation has run its full course and an impasse is inevitable (Stipanowich, 1996b).

The benefits of mediation are well documented. A joint study by Cornell University, the Foundation for the Prevention and Early Resolution of Conflict (PERC), and Price Waterhouse LLP revealed that 90% of respondents viewed mediation as a cost-saving measure for their corporation, and 66% of respondents said mediation provides more "satisfactory settlements" (Lipsky and Seeber, 1997).

Similar results were found in an American Arbitration Association study (American Arbitration Association, 2003). Drawing on a pool of 101 Fortune 1000 companies, 103 mid-size public companies, and 50 privately held companies, the AAA study found that 91% of respondents used mediation because it saved money; 84% used mediation because it saved time; 61% said mediation provides more satisfactory settlements (American Arbitration Association, 2003, p. 24). Additionally, 77% of respondents believed mediation saved costs in comparison to litigation, while only 58% of respondents felt the same way when comparing arbitration and litigation (American Arbitration Association, 2003, p. 19).

## *Dispute review boards*

Dispute Review Boards (DRBs) are a mixture of both expert opinions and neutral evaluations. Initially developed within the construction industry, DRBs are substantiated in industry experience and have been in use for more than 40 years. The first recognized project to utilize a DRB was the Eisenhower Tunnel in Colorado and during the first 10 years of its existence, DRBs were primarily used on tunnel and heavy civil projects (Matyas *et al.*, 1996). However, today its application in other construction projects is becoming more accepted (Dispute Resolution Board Foundation, 2019).

The basic structure of a DRB consists of a three-member, expert panel appointed by both the owner and the contractor. The critical difference between a DRB and other forms of alternative dispute resolution is that the DRB team meets both before and during construction operations. This allows the DRB members to familiarize themselves with the people, process, and project specifics. Using project drawings, specifications, and site visits, the DRB makes non-binding recommendations to the parties who cannot resolve issues at the project level (Dispute Resolution Board Foundation, 2019). "Acceptance by the parties is facilitated by their confidence in the DRB — in its members' technical expertise, firsthand understanding of the project conditions, and practical judgment — as well as by the parties' opportunities to be heard (Matyas *et al.*, 1996, p. 3)."

An important organization dedicated to the increasing use of DRBs is the Dispute Resolution Board Foundation (DRBF). DBRF is a non-profit organization that encourages the avoidance and resolution of disputes through the application of DRBs. "The Foundation provides assistance with the worldwide application of the DRB method by providing general advice and suggestions tailored for the conditions and practices existing in local areas (Dispute Resolution Board Foundation, 2005)." According to DRBF data from a 2013 Florida Department of Transportation study, projects adopting DRBs in the contract documents incurred 17.4% less time and 5.8% less cost overruns than projects without DRBs (Dispute Resolution Board Foundation, 2019).

The DRBF indicates, "DRB costs range from 0.05 percent of final construction contract cost, for relatively dispute-free projects, to a maximum of 0.25 percent for difficult projects with disputes. [The average was] 0.15 percent of final construction contract cost, including an average of four dispute recommendations (Dispute Resolution Board Foundation, 2005)."

## Resolving disputes off the jobsite

If disputes continue to be unresolved after attempts to resolve them on the jobsite fail, additional mechanisms must be explored to address the situation. Resorting to systems and procedures that take resolution efforts off the jobsite mark a significant shift in the amount of time, cost, and emotional effort that must be expended (Alaloul *et al.*, 2019; Richter, 2000). The following sections address these alternatives.

### *Neutral evaluation*

A neutral evaluation is a private and non-binding technique whereby a third party, usually legally qualified, gives an opinion on the likely outcome at trial for the basis of settlement discussions (Office of Government Commerce, 2002). An example of neutral evaluation is mini-trials. Groton (1997, p. 55) defines the mini-trial as, "A brief presentation of each side's 'best case' arguments in the presence of principle executives of both parties, whose efforts to settle the dispute are usually facilitated by a neutral."

Mini-trials are conducted as if the case were being presented in front of a jury or arbitration panel. Parties make a presentation to a neutral third party and a panel of senior executives from each side (Alaloul *et al.*, 2019). In order to make an effective summary presentation, each side must have a reasonable amount of time for limited discovery so that they gain an understanding of their own contentions as well as the opposing side's contentions (Hinchey and Schor, 2002). The third-party neutral is responsible for providing a thorough assessment to both parties of the dispute. The disputing parties can then decide to settle based on the neutral's evaluation or move to another form of resolution (Mays, 2003).

## *Adjudication/expert determination*

Expert determination is defined as, "a private process involving an independent expert with inquisitorial powers who gives a binding decision (Office of Government Commerce, 2002, p. 4)." This term is most frequently encountered in the construction industry in the UK. Adjudication is a binding decision made by an appointed neutral, often a quantity surveyor, either by deciding on the basis of submitted documents, or as is increasingly the case, after a hearing. It is designed to provide a speedy, if not always elegant, resolution to enable work to continue on-site without interruption (McGeorge *et al.*, 2007). Either party may appeal the adjudicator's decision to court or arbitration or indeed settle the dispute by mediation. The Housing, Grants, Regeneration Act 1996 in the United Kingdom has greatly increased the use of adjudication (Brooker and Lavers, 1997; Office of Government Commerce, 2002).

## *Arbitration*

Arbitration is defined by the American Arbitration Association (AAA) as, "the submission of a dispute to one or more impartial persons for a final and binding decision. Through contractual provisions, the parties may control the range of issues to be resolved, the scope of relief to be awarded, and many procedural aspects of the process (American Arbitration Association, 2004)." In the United States, arbitration has been the dispute resolution method of choice for many years in the construction industry and is included in many industry standard contract documents including the American Institute of Architects and the Associated General Contractors of America (Alaloul *et al.*, 2019; McGeorge *et al.*, 2007).

Typically, the proceedings are administered by an organization, such as the AAA, which will have specific rules for the process (Nelson, 2003). The AAA Construction Industry Arbitration Rules and Mediation Procedures is a lengthy document covering every detail of the proposed arbitration procedures for the construction industry. However, parties may set up their own rules in the contract as long as both parties agree with those rules. Final decisions of the arbiter are final, binding, and generally not reviewable by the court system (King *et al.*, 2019). A detailed coverage of arbitration usage in construction can be found in Stipanowich (1987).

Despite centuries of use in resolving disputes (Stipanowich, 1996b), arbitration has recently received sharp criticism from academics and practitioners alike. Anecdotes about the process, the arbitrators, and the decisions have shown that arbitration proceedings are becoming more and more like litigation (Cakmak, 2021; Cheung *et al.*, 2002; Harmon, 2003; Keil, 1999; Reuben, 1996; Stipanowich and O'Neal, 1995). Braun (1998, p. 9) writes, "Arbitration can be an expensive, unending kangaroo court in which the concepts of justice and fairness are trampled and neither the arbitrators nor the arbitration association seems to have any interest in anything other than maximizing the fees paid to them by the parties." In addition to the condemnation of the high costs of arbitration procedures, many critics argue that arbitration is also an unnecessarily lengthy process (Lurie, 2005). Despite these criticisms, arbitration remains a standard part of many standard forms of contract.

## A comprehensive dispute management approach

As previously discussed, disputes should try to be prevented from occurring on projects, but that is not always possible. Creating a comprehensive management approach that allows parties to resolve disputes at the lowest level possible and without ligation is highly desirable. Whether using negotiation, mediation, arbitration, dispute review boards, or any other means for resolving disputes, the main purpose is to reach equitable solutions quickly and with as little distraction as possible. Unfortunately, the overabundance of options available to today's industry practitioners for resolving disputes can be both confusing and time consuming (Gebken, 2006). Figure 3 shows the typical process for a dispute to be

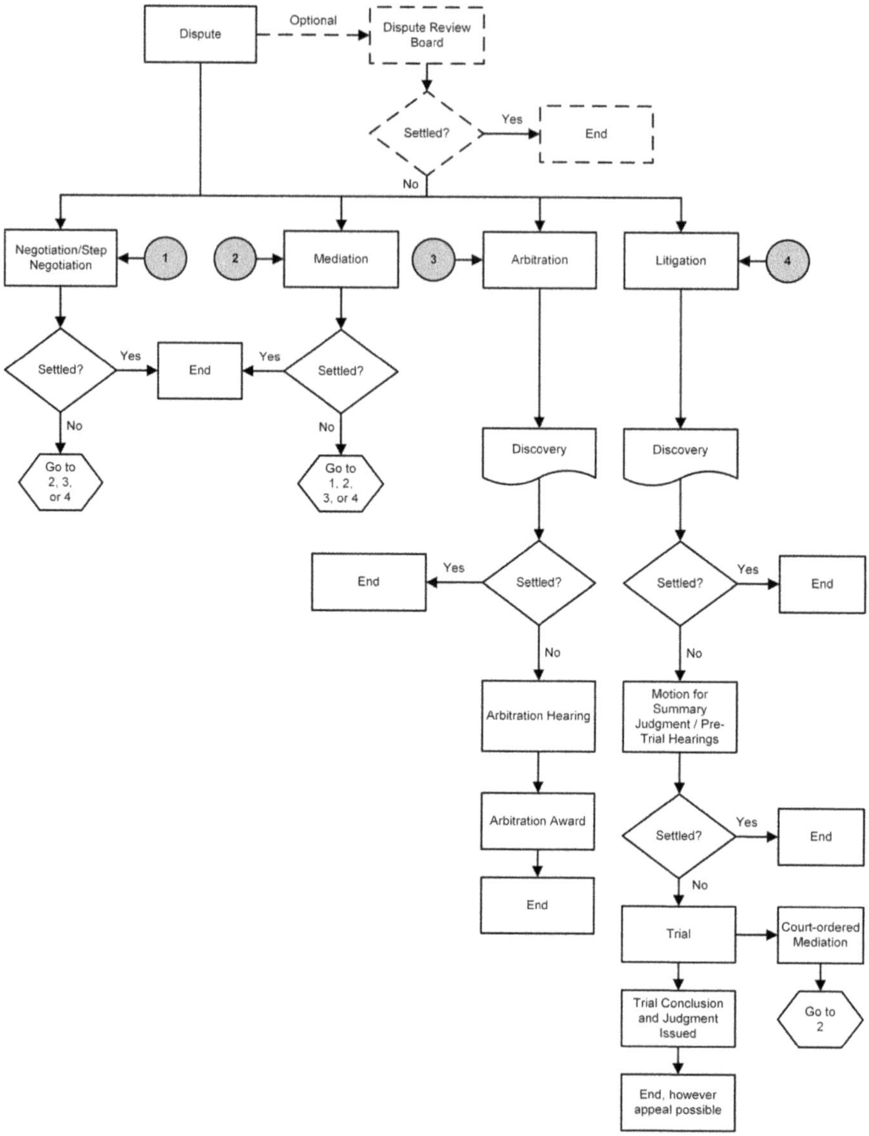


**Figure 3:** Typical dispute process diagram.

resolved in the construction industry. This complex process diagram validates two key ideas. First, disputes are an inevitable facet of construction projects. Second, no one method of performing dispute resolution works in every instance. The key is to understand the proper application and associated benefits of each system in the context of specific project environments.

As each dispute is unique to the situation from which it developed, it is necessary to build dispute resolution systems that are not only flexible but also adaptive to the particulars of the dispute. An example of such a system is the Flexible Framework for the Prevention and Resolution of Construction Disputes (Pappas, 2004). This system creates a dispute resolution advocate who is responsible for assisting parties to adopt dispute resolution procedures that are most appropriate for the subject matter in dispute. It has been the observation of the author that many disputes are not resolved at the early stages because the parties feel as if they are merely stepping stones along the path toward an "authoritative" conclusion of the matter. Some parties treat negotiations and even early mediation efforts as positional bargaining or tactical maneuvering tools for settlement later.

While step negotiations and dispute resolution systems based upon an escalating progression of ADR tools make logical sense in a broad view, it is the forced nature of pursuing one option before another and another that builds hostilities and additional costs into the resolution efforts. A better approach to dispute resolution designs is to focus on directing disputes toward a resolution procedure that will assist parties in resolving differences as quickly and as inexpensively as possible.

## Discussion

### *Construction dispute trends*

Studies into the trends of construction disputes have been inconsistent at best. An Engineering News Record editorial states, "...we still don't know definitively whether we [the construction industry] have had any success combating lawsuits. All we have is 'possibly' and 'presumably' (Editorials, 1997, p. 62)." An Associated General Contractors/Deloitte Touche LLP survey reported that the cost of litigation had become less significant for general and specialty contractors since 1994 (Flanigan *et al.*, 1997). More recently, Arcadis published its 10th annual Global Construction Disputes Report where they found that the global volume of disputes increased and the average value and the time taken to resolve them decreased (Arcadis, 2020).

Two noteworthy studies examining the trends in construction industry disputes were performed on both the US Naval Facilities Engineering Command (NAVFAC) and the US Army Corps of Engineers (USACE). In the NAVFAC study (Kilian and Gibson, 2005), the authors found a trend toward the reduction of litigation over the past 20 years. The USACE study (Kurgan, 2005) found a similar decrease in the number of construction claims given a fairly constant construction budget throughout the study period. Both the NAVFAC and the USACE studies reveal a significant reduction in the amount of construction-related litigation in the study years before and after 1993 — the time period where the emergence of design–build and partnering practices in NAVFAC and USACE construction contracts occurred (Kilian and Gibson, 2005; Kurgan, 2005). The overall downward trend may be due to a number of factors including the successful implementation of partnering, the more frequent awarding of design–build and cost plus contracts, best-value selection, and a possible paradigm shift in internal policy on the part of NAVFAC and USACE toward its claim settlement process. While Kilian (2003) and Kurgan (2005) found nothing to confirm or refute these observations, it stands to reason that the use of partnering and design–build would lower the instances of litigation as they both provide an opportunity for improved communication and problem solving based upon intuitive reasoning.

A similar conclusion was drawn in the Associated General Contractors/ Deloitte Touche LLP survey which found that the perceived overall significance of litigation had decreased over a six-year period (Flanigan, 2000). The cause was attributed to both the familiarity with and use of ADR techniques by both General and Specialty Contractors and the effectiveness of partnering and quality programs. Again, no specific causal link between these perception changes was able to be drawn.

Arcadis' study also cites collaboration between project participants as a key factor in reducing dispute occurrence. The collaborative or cooperative behaviors that were identified as most important in the mitigation or early resolution of disputes include the following: (1) owner/contractor willingness to compromise; (2) accurate and timely schedules and reviews by project staff or third parties; (3) contract and specifications reviews; (4) proactive risk management (Arcadis, 2020).

The key takeaway needs to be that project teams should be thinking about disputes on a holistic approach to ensure successful outcomes are achieved. To do this, dispute risks must be known and accounted for. The next section will explore how risk management philosophy can integrate into a comprehensive approach to disputes.

## *Dispute risk management*

Because of the potentially large financial impacts of disputes in the industry, a savvy construction practitioner should attempt to limit possible consequences in the same manner they would address other construction risks. Construction disputes should, as outlined in the risk management literature, be identified, assessed, and controlled.

While there has been a lot of literature covering disputes, the focus has been on discussing dispute identification, dispute control, and to some extent dispute frequency within the industry, so much less attention has been paid to calculating dispute severity/costs. For industry practitioners to continue their efforts of preventing disputes and minimizing their impacts, they must not only improve their ability to estimate the true impacts of disputes but also adopt a standardized risk management framework. Figure 4 illustrates an example of how these risk management concepts can be implemented for dispute prevention and resolution.

Based upon the discussion throughout this chapter, nearly every component of the dispute risk management framework has been explored and expanded upon to help industry professionals deal with inevitable conflict. The causes of construction conflict and tools to proactively identify and predict their possibility were both presented. Methods to control and resolve disputes through collaborative behaviors and ADR procedures were documented from multiple sources. Ultimately, it is the severity piece that needs more attention and data. Industry participants must strive to collect as much data and information as possible concerning disputes in order to establish some benchmarks for future improvement. When the costs of resolving disputes are known, companies can evaluate their dispute history and see how money spent on transactional costs of resolution might be better spent on preventative measures. Owners, contractors, and subcontractors alike can review their contract documents and tailor the

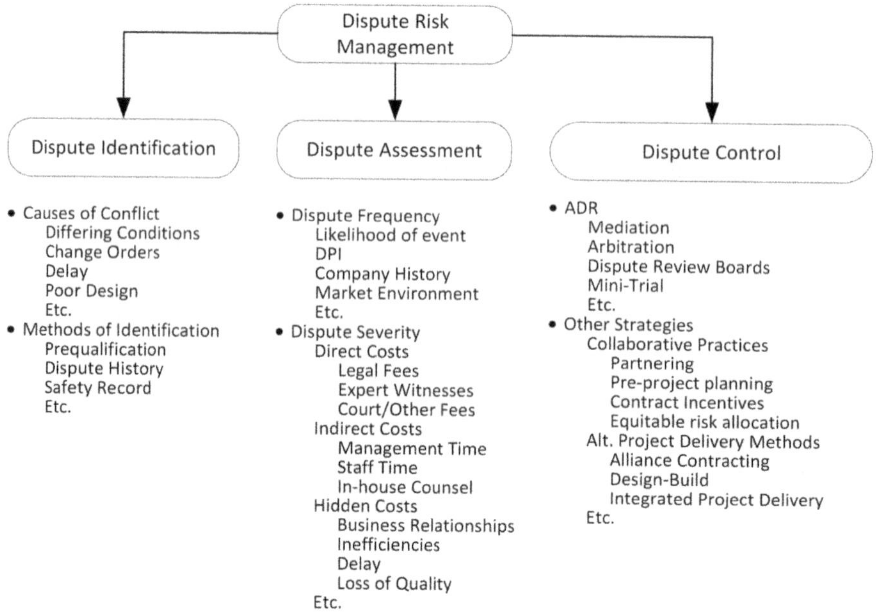


**Figure 4:** Dispute risk management framework.

language to each individual project depending on the comprehensive dispute management system established by each organization.

Fully adopting a dispute risk management approach is a critical step toward further reducing the impacts of disputes on the construction industry. The potential for economic savings is enormous.

## Conclusions

"Discourage Litigation. Persuade your neighbors to compromise whenever you can. Point out to them how the nominal winner is often a real loser — in fees, expenses, and waste of time (Lincoln, 1850)." Abraham Lincoln's famous 1850 quote on the pitfalls of litigation rings as true today as it did more than 170 years ago. This advice is especially pertinent for the construction industry as one industry expert cited that nearly \$5 billion USD is spent on construction litigation alone each year and that this number increases 10% each year (Michel, 1998). If this is the cost for merely construction litigation, imagine how much is expended on resolving disputes as a whole, since 95% of disputes never see the inside of a courtroom (Stipanowich, 2004).

As conflicts are inevitable, their destructive component, project disputes, must be reduced if not prevented altogether. Parties must work to establish dispute management systems that address both prevention and minimization, the latter of which should seek to resolve conflict at the lowest levels possible with personnel directly involved in the project. Likewise, the resolution effort and the people involved in its implementation should also be directly aware of the project situation as well. Dispute review boards have proven that keeping a knowledgeable third-party agent up to speed with project progression and "on-call" to advice upon conflict situations is an investment that pays dividends in multiple areas (Dispute Resolution Board Foundation, 2019).

However, resolving a dispute after a project has been completed with lawyers, judges/arbitrators, and jury members that are far removed from the actual project costs everyone money, time, and much aggravation. Figure 5 reexamines Figure 2 and modifies some of the relative cost of dispute resolution efforts based upon development in industry.

In the revised model, the relative costs of both mediation and binding arbitration rise well above and potentially closer to the costs incurred

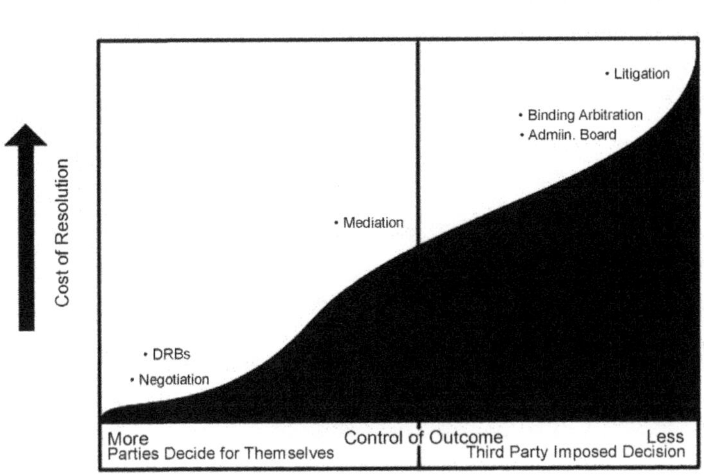


**Figure 5:** Revised control of outcome vs. cost of dispute resolution.

*Source*: Adapted from Richter, 2000.

during full trial litigation. While negotiation remains low on the cost curve, the costs of this ADR procedure may be closer to the costs of implementing a DRB than previously thought. Lastly, the overall slope of the curve has also been adjusted to account for the rapid escalation of resolution costs beyond a party-to-party negotiation. Vorster (1993) called the difference between resolution methods on and off of the jobsite as "the continental divide" of dispute costs.

The industry must continue to be vigilant about understanding and managing the risks of construction disputes. Additional data points on dispute decision-making will only improve the comprehensive model of dispute risks management in practice. Exploring how settlement values are affected by estimated transactional costs or if company backlog/revenue volumes impact dispute resolution efforts are two examples of where developments are still needed. Enabling industry professionals to make informed decisions with the most accurate data available is the reason as to why these data are so important.

The overriding conclusion must be that change, and equally conflict, is inevitable. As such, mechanisms or systems that can be established a priori can help reduce the costs of resolving problems on the construction site. Additionally, these in-place systems can greatly reduce the hidden costs of injured business relationships, tarnished corporate images, and anguished personal lives by having a system that can be easily followed in times of uncertainty.

# About the Author

**Richard J. Gebken** is an Associate Professor and Department Head in the Department of Technology and Construction Management at Missouri State University where he holds the McBride & Son Homes Professorship and serves as the Construction Management Program Coordinator. His research interests include construction law, alternative dispute resolution, and pedagogical inquiry. Dr. Gebken teaches construction cost management, electrical systems for buildings, construction law, and project administration.

# Chapter 15

# Proposed Framework and Building Blocks for a Body of Knowledge in Construction Project Delivery, Procurement, and Contracting

Pramen P. Shrestha[1] and Giovanni C. Migliaccio[2]

[1]*University of Nevada Las Vegas*
[2]*University of Washington*

This book includes contributions by multiple scholars active in studying the contiguous knowledge areas of project delivery, construction procurement, and contracting. By now, the editor can rely on these contributions to outline a series of building blocks for a body of knowledge on these knowledge areas.

## Section A: Project delivery

The first section of this book focuses on *project delivery* with a focus on project delivery methods used in the construction of vertical and horizontal projects. Out of the five chapters, the first chapter covers a brief description of the construction project phases, the generally accepted definition of the *project delivery method*, various types of delivery methods, and the method of financing a project by public agencies using a public–private partnership (PPP) contract model. As "one size does not fit all", one delivery method is not always suitable for all construction projects. The owner needs to consider various aspects of the project while choosing the right delivery method for their project. The literature includes significant contributions that identified the critical success factors for choosing the delivery method for various types of projects. The studies have found various advantages and disadvantages of each delivery method. With its simplicity, design–bid–build (DBB) has been the prevalent approach to delivering projects and is based on separate contracting design and construction services from designers and contractors; this method has been successfully implemented by various public agencies and private organizations. After being abandoned during the initial part of the last century, the design–build (DB) method has been used extensively for building projects and for the last three decades in highway construction projects and other infrastructures. In the beginning, lump-sum DB was only used in infrastructure projects; however, owners have started using progressive DB methods over the last decade. Similarly, the construction-manager-at-risk (CMAR) method has been extensively used in residential and commercial building projects. One of the major advantages of the DBB method is the lowest bid price; however, it is not confirmed that the lowest bid price will turn out to be the lowest cost for the project after its completion. The DB method has the advantage of completing the project faster compared to DBB. CMAR has a distinct feature that helps the owners bring the contractors on board during the design phase, which can help the constructability issue on the project as well as conduct value engineering to reduce the cost of the project. As the public agencies are struggling with financing the projects, the PPP method helps them bring the private parties to the project for funding. The private parties can also perform the design, construction, operation, and maintenance of the projects.

One of the important factors in using a delivery method for the design and construction of the project is how to integrate the recently available digital technologies during the design, construction, and operations of the capital projects. To solve this problem, the concept of *Integrated Digital Delivery (IDD)* has emerged recently. IDD consists of the digital representation of capital — buildings or infrastructure — projects. The main goal of IDD is to create the digital models during the planning and design phase of the projects and use these models during the construction and operations of the physical assets. This often relies upon the development of Building Information Models (BIM) for building projects or Civil Infrastructure Models for infrastructure projects during their design, construction, and maintenance during operations. These digital models of the physical capital project and its processes are also known as digital twins.

The advantages of IDD are that, when the changes are incorporated in the model, it will be easier to calculate the changes in the quantities as well as plan the construction activities accordingly so that it will be safe and cost-effective. These digital models can also be used as as-built drawings to conduct the maintenance of the physical facilities during the operation phase. This is very helpful for the asset managers who are responsible for the maintenance and operations of the building or infrastructure projects. As IDD is new, there are many barriers, including the cost of adopting technologies, social resistance, and status quo. However, with the advancement in technology and the addition of new workforce in the industry, IDD is initiating a new trend in the industry while also supporting the ongoing shift toward integration of design, construction, operation, and maintenance of building or infrastructure projects.

One of the important aspects of the success of any project delivery method depends upon the relationship between the various project participants (owners, designers, contractors, or financiers) and the work environment of the project. The *relationships and the work environment* in the projects that use various types of delivery methods differ in their contractual relationship and the collaboration deemed necessary by this relationship. All the delivery methods alternative to DBB, including DB, CMAR, and PPP, are also called collaborative delivery methods because the collaboration among owners, designers, and the contractors is high compared to DBB. As reported in some chapters of this book, research has found that the DBB is arranged in such a way that it creates adversarial relationships between the project participants because it requires individual project participants to work independently without the collaborative environment in the project. In case of DB, the designer and the contractor are under one entity; therefore, the collaboration among these two parties is very frequent. In lump-sum DB, the feedback mechanism from the owner during the design phase is minimal, and the owner does not have control over the design of the project. Therefore, the improved version of the lump-sum DB has been recently introduced as progressive DB. In progressive DB, the feedback mechanism over the design of the project from the owner has been included. Therefore, the owner, designer, and contractor work together to design the project in such a way that the constructability, value engineering, and operation issues can be integrated collaboratively during the design phase. The CMAR method highly integrates the collaborative environment by bringing together all of the parties involved in the projects during the design and construction phases. This method also allocates risks to the parties who can handle the risk cost effectively. This delivery method allows hand-picking all of the team members and, therefore, naturally enables much more collaboration. As all the PPP projects follow the DB method to design and build the projects, the collaboration and work environment are similar to that of regular DB projects.

The *contractor's involvement early in the project phase* has been identified as a key driver to the success of any construction project. The alternative contracting methods like DB, CMAR, and PPP have early contractor involvement either in their design phase or from the inception phase. However, the traditional delivery method DBB never allows the contractor to be involved until the bidding phase of the project. Various studies have shown that the design without the input of the contractors will not be feasible to build and will meet various constructability issues. However, the owners using DBB methods have recently used various processes, e.g., Value Engineering (VE) or Alternative Technical Concept (ATC), during their construction bid phase to bring innovation in design as well as reduce the cost of the project. VE brings the innovation in design so that the alternative design can be proposed without compromising the functions of the facilities and reduce the overall costs (construction, operation, and maintenance) of the facilities. This can be conducted during the design phase or immediately after the design is completed. ATC is a new method used in US highway construction in which the owners will ask contractors to propose the alternative design during the construction bidding phase. The owner will review the alternative technical concepts and will incorporate them if deemed necessary during the contract execution phase. ATC can also be included in DB and CMAR projects. In CMAR projects, the ATC is incorporated before the design contract is awarded; however, in DB projects, the ATC will be incorporated during the bid phases. There are various barriers to bringing the contractors into the projects early, including, but not limited to, organizational staff, legal and contractual issues, and work culture differences. To overcome these barriers, the owners need to implement these processes during the project development phase.

As mentioned before, the delivery method used in different types of projects can be dependent upon various factors. One of the chapters reports research findings on the *preference of delivery method in the construction of green buildings*. Design–build was found to be the owner-preferred method for projects that incorporate sustainability among their objectives. Green buildings are complex in nature and need to integrate various technologies and multidisciplinary teams in order to conform to their required specifications. As DB promotes integration as well as value on collaborative work environments, the majority of green building owners prefer the DB delivery method to complete these types of projects. It has also been found that the green buildings delivered through the DB method had significantly higher sustainability performance outcomes compared to the green buildings delivered through DBB and CMAR.

The significance of the project delivery selection method, integrated digital delivery, collaboration and work environment, early contractor involvement, and use of project delivery methods in the green building construction have been highlighted in the initial section of this book. These are some of the important issues the readers need to consider while selecting the delivery method appropriate for their future projects. The timing of the delivery method selection, the analysis of critical success factors, the risks analysis, and the allocation process are key to the success of any construction project. The initial section of the book has educated the readers on various aspects of the delivery methods so that they can choose the appropriate delivery method for the project to be successfully delivered on time, within budget, and with quality and safety.

## Section B: Procurement

In the second section, various aspects and *approaches to construction procurement* are described; the description often includes a discussion on their strengths and weaknesses. Described approaches include sole sourcing, low bidding, as well as qualifications-driven and best-value procurement processes; this chapter also analyzes methods for incentivizing sustainability in public procurement.

One of the methods to choose the designers or contractors in any construction project is to use *qualifications-based selection* criteria during the selection process. In design–bid–build projects, this method is generally used to select the designers, as it is difficult to determine the bid of the design projects during the bidding process. Therefore, the public agencies select the designers based on their qualifications and the work experience they have gained in that area of design. However, it will be noted that the owners need to refer to their federal, state, or local procurement rules before using this procurement method. The qualifications-based procurement selects the construction contractors not only based on their bid cost but also their qualifications during the selection phase. The contractors in DB and the construction manager in CMAR projects are generally selected based on this procurement method. The owner can use this method to select contractors using either the "one-step" or "two-step" process. The former method consists of issuing the request of the proposal to the contractors in which the contractors' qualifications and cost components will be included. The latter method first invites the contractors to submit their qualifications and, based on these qualifications, the limited contractors will be prequalified. Then in the second step, the owner will ask the prequalified contractors to submit a full technical proposal along with a bid cost so that the owners can select the contractor based on their qualifications and the cost. The owner needs to choose one of the above methods based on their project needs and other relevant factors.

As previously stated, the DB project delivery method has been adopted only in the last few decades for infrastructure and highway projects specifically. For these projects, the DB method has been coupled with best-value procurement and lump-sum contracting. Under *best-value procurement*, design-builders are selected through award algorithms that combine multiple criteria, including cost and non-cost criteria, which are weighted individually. The goal is to select the offer that would provide the best value to the client. One of the chapters reports research findings on the effectiveness of these processes to identify the best value offer and balance the selection around value for highway projects. The study found that agencies rely on best-value procurement to incorporate all relevant factors in the selection of the design–build entity, including price and non-price factors. However, the nature of the algorithm, including the criteria selected, the weights assigned to them, and the scoring procedures often result in a selection decision strongly biased by price and whose results mirror that of a low bidding approach.

As sustainable outcomes of capital projects have become more valuable, it is important to *adapt the selection of built environment service providers to address the owner's desires for sustainability*. Procurement in the private sector is more fluid and lacks the restrains of public procurement, which allows owners to custom fit the selection of service providers around their project objectives, either inclusive or not of sustainability goals. On the other hand, adapting public procurement to these new project objectives has often encountered technical and organizational challenges. One of the chapters provides a practice-grounded examination of the technical and organizational challenges in integrating sustainable practices into public procurement, identifies the best practices, and outlines the underlying principles for incorporating sustainability into public procurement and its policies with a focus on the highway sector. Due to its multi-criteria selection approach, best-value procurement is identified as the most promising procurement route for sustainable outcomes. Furthermore, specific processes — such as alternative technical concepts, life-cycle assessment-based approaches, contractual instruments and specifications, and green rating systems — are dissected and analyzed for their ability to select providers based on the expected sustainable outcomes. Moreover, an overview of how procurement policies can incorporate, mandate, or promote these processes is shared with the readers.

## Section C: Contracting

In the third and last section, readers have been provided an overview of various approaches to construction contracting. In addition to an initial overview of the most traditionally used contracting approaches, performance-based contracting, contractual aspects of integrated project delivery, and international contracting are offered. As project delivery, procurement, and contracting do not always succeed, projects are often plagued by disputes, which are discussed in this section as well. Out of the five chapters, Chapter 10 provides an overview of contracting and describes the most *traditional contracting approaches*. Contracting is a process that often initiates during procurement through the drafting of contract documents and is completed once a contract is signed by the parties, which provides them with instructions on how to succeed on the project. Improper instructions are one of the leading causes of construction disputes. In addition to the contractual agreement, a series of other contract documents are paramount to contracting, including general conditions, special conditions, construction drawings, and construction specifications. As the project delivery method would strongly affect the instructions included in the contract documents, readers are provided information on the statutory framework regarding construction contracts under various delivery methods as well as various standard forms of agreement as developed and curated by a large number of industry associations. Finally, a classification of contracts by parties involved and a pricing method has been provided.

Traditional contracting relies on a method-based contracting (MBC) approach (also known as prescriptive-based contracting), which had served well the traditional form of project delivery, DBB. As alternative forms of delivery have been introduced to practice, contracting has been adapted. A major adaptation has been the development and introduction of *performance-based contracting (PBC)* as an alternative or complement to MBC. Differently from MBC, which provides a set of instructions on the method to achieve project objectives, PBC focuses on describing the expected end result while leaving some freedom to the contractor on how to achieve it. As this approach may result in a variety of results, contracting agencies often provide incentives and disincentives based on the contractor's ability to satisfy the project objectives. PBC substantially differs upon how performance is measured, the type of project, and the type of assets to be subject to performance-based instructions in the contract documents. PBC has been used worldwide to various extents, and readers have been exposed to an overview of the history, geographical adoption, and approaches for PBC.

Among all alternative forms of project delivery, *integrated project delivery* (IPD) is the most recent and revolutionary. IPD introduces a radical shift in contracting through the use of multiparty agreements, which allow owners to infuse the project environment with partnering concepts and processes. While partnering is a management practice that has been used in all forms of project delivery and is sometimes made a contractual requirement, IPD projects rely on partnering as the foundation for their delivery and contracting. This has substantial implications on various aspects of contracting, including the following: (a) the number of parties signatory of the contract, a more immersive involvement in contract management by all parties; (b) a joint agreement on the target cost, a continuous reliance upon value engineering; (c) a focus on aligning interests and objectives of all project parties, a consequent reduction of conflict of interest; (d) a transparent sharing of information through both traditional and digital forms, such as BIM; (e) a shared approach to risk management, an avoidance of claims and disputes through specific contract clauses; (f) a performance-based evaluation tied to incentives and, sometimes, disincentives; (g) a collaborative project organization culture. While IPD offers many benefits that have been described in Chapter 12 and documented in the literature, challenges also exist for its implementation, which were presented.

As the construction supply chain has become global and the frequency of large projects has increased, construction firms have seen an increasing need in working with foreign entities or delivering construction projects in other jurisdictions across country borders. The readers have learned about *International Contracting* (IC) as a business model in that an AEC organization contracts for and performs cross-border construction projects, including design and construction services, the supply of materials and equipment, and increasingly in recent years, financing, and operation and maintenance (OM) services, in return for revenues and profits. In addition, Chapter 13 provided a snapshot of the current status of the international contracting market, including specific project delivery methods and contracts.

As project delivery, procurement, and contracting do not always succeed, projects are often plagued by disputes. Chapter 14 has provided an extensive overview of *construction disputes*. As disputes are originated by claims and/or conflicts on projects, the chapter begins describing the relationship between these events before identifying main sources of construction disputes and defining the concepts of likelihood, impact, and severity of disputes. Still, a description of the main dispute resolution approaches, both at and off the jobsite level, is provided. The chapter later focuses on moving toward dispute prevention, minimization, and resolution by outlining a comprehensive dispute management approach.

# About the Authors

**Pramen P. Shrestha** is a Professor of Construction Engineering Management and Graduate Coordinator in the Civil & Environmental Engineering and Construction department at the University of Nevada Las Vegas (UNLV). He holds an M.Sc. in Civil Engineering from Oklahoma State University and a Ph.D. in Civil Engineering from the University of Texas at Austin, United States (US). He completed his B.Sc. in Civil Engineering from the National Institute of Technology, Jamshedpur, India. He had about 10 years of professional experience in design and construction of highway and infrastructure projects in Nepal before moving to the US. After completing his Ph.D. in Civil Engineering from the University of Texas at Austin in 2007, he joined UNLV as an Assistant Professor. He has received federal, state, and private research grants worth about US $2.5 million. His research grants are funded by the US State Departments of Transportation, Transportation Research Board, National Science Foundation, US Labor Department, Water Design Build Council (WDBC), Airport Cooperative Research Program (ACRP) and private agencies. He has helped with the graduation of five Ph.D. and 34 master's students as an advisor. He currently has two Ph.D. students and three master's students working for him on various projects. He has published more than 55 peer-reviewed journal papers and 75 conference proceedings. He is a Fellow of American Society of Civil Engineers and a licensed professional engineer in Texas.

**Giovanni C. Migliaccio** holds a Howard S. Wright Endowed Professorship, he is the current Chair of the Department of Construction Management, and he has served as the Executive Director of the Center for Education and Research in Construction (CERC). He holds an M.S. and a Ph.D. in Civil Engineering from the University of Texas at Austin and a master-level degree from Politecnico di Bari in Italy. Prior to moving to the US, he worked in Italy in the construction management of telecommunication projects under Nortel Networks, Nokia Networks, and IPSE 2000. At UW, Giovanni is active in all three pillars of academic life, including teaching, research, and service to the university and the industry.

# References

## Chapter 2

Abdul-Aziz, A.-R. (1994). Global strategies: A comparison between Japanese and American construction firms. *Construction Management and Economics*, 12(6), 473–484. https://doi.org/10.1080/01446199400000059.

Acquisti, A., Adjerid, I., Balebako, R., Brandimarte, L., Cranor, L. F., Komanduri, S., Leon, P. G., Sadeh, N., Schaub, F., Sleeper, M., Wang, Y., and Wilson, S. (2017). Nudges for privacy and security: Understanding and assisting users' choices online. *ACM Computing Surveys*, 50(3), pp 1–41. https://doi.org/10.1145/3054926.

Adriaanse, A., Voordijk, H., and Dewulf, G. (2010). The use of interorganisational ICT in United States construction projects. *Automation in Construction*, 19(1), 73–83. https://doi.org/10.1016/j.autcon.2009.09.004.

Afuah, A. N. and Bahram, N. (1995). The hypercube of innovation. *Research Policy*, 24(1), 51–76. https://doi.org/10.1016/0048-7333(93)00749-J.

Alreshidi, E., Mourshed, M., and Rezgui, Y. (2018). Requirements for cloud-based BIM governance solutions to facilitate team collaboration in construction projects. *Requirements Engineering*, 23(1), 1–31. https://doi.org/10.1007/s00766-016-0254-6.

American Institute of Architects. (2007). Integrated Project Delivery. http://content.aia.org/sites/default/files/2017-02/Integrated Project Delivery Guide.pdf.

Arayici, Y., Onyenobi, T., and Egbu, C. (2012). Building Information Modelling (BIM) for Facilities Management (FM). *International Journal of 3-D Information Modeling*, 1(1), 55–73. https://doi.org/10.4018/ij3dim.2012010104.

Azhar, S. (2011). Building Information Modeling (BIM): Trends, benefits, risks, and challenges for the AEC industry. *Leadership and Management in Engineering*, 11(3), 241–252. https://doi.org/10.1061/(asce)lm.1943-5630.0000127.

Belle, I. (2017). The architecture, engineering and construction industry and blockchain technology. *2017 National Conference on Digital Technologies in Architectural Education and DADA 2017 International Conference on Digital Architecture*, pp. 279–284. Nanjing, China: Nanjing University.

Blayse, A. M. and Manley, K. (2004). Key influences on construction innovation. *Construction Innovation*, 4(3), 143–154. https://doi.org/10.1108/14714170410815060.

Borrmann, A., König, M., Koch, C., and Beetz, J. (2018). Building information modeling: Technology foundations and industry practice. In *Building Information Modeling: Technology Foundations and Industry Practice*. Cham, Switzerland: Springer. https://doi.org/10.1007/978-3-319-92862-3.

Building and Construction Authority. (2018). Integrated Digital Delivery (IDD). https://www.bca.gov.sg/IntegratedDigitalDelivery/IDD_benefits.html.

Building and Construction Authority and Infocomm Media Development Authority. (2018). BCA-IMDA Call for submission — Construction Digital Platforms (CDP). https://www.imda.gov.sg/-/media/imda/files/industry-development/infrastructure/digital-platforms/bca-imda-call-for-submission-construction-digital-platform-briefing-slides.pdf?la=en.

Chan, A. P. C., Darko, A., and Ameyaw, E. E. (2017). Strategies for promoting green building technologies adoption in the construction industry — an international study. *Sustainability*, 9(6), 969–986. https://doi.org/10.3390/su9060969.

Costa, A. A. and Grilo, A. (2015). BIM-based e-procurement: An innovative approach to construction e-procurement. *Scientific World Journal*, 1–15. https://doi.org/10.1155/2015/905390.

Dainty, A. R. J., Briscoe, G. H., and Millett, S. J. (2001). Subcontractor perspectives on supply chain alliances. *Construction Management and Economics*, 19(8), 841–848. https://doi.org/10.1080/01446190110089727.

Darko, A. and Chan, A. P. C. (2017). Review of barriers to green building adoption. *Sustainable Development*, 25(3), 167–179. https://doi.org/10.1002/sd.1651.

Djokoto, S. D., Dadzie, J., and Ohemeng-Ababio, E. (2014). Barriers to sustainable construction in the Ghanaian construction industry: Consultants perspectives. *Journal of Sustainable Development*, 7(1), pp. 134–143. https://doi.org/10.5539/jsd.v7n1p13++4.

Eggers, W. D., Turley, M., and Kishnani, P. (2018, June 19). Soft law: New tools for governing emerging technologies. https://www2.deloitte.com/insights/us/en/industry/public-sector/future-of-regulation/regulating-emerging-technology.html.

Enshassi, A., Abu Hamra, L., and Mohamed, S. (2016). Barriers to implementation of Building Information Modelling (BIM) in the Palestinian construction industry. *International Journal of Construction Project Management*, 8(2), 103–123.

Froese, T. M. (2010). The impact of emerging information technology on project management for construction. *Automation in Construction*, 19(5), 531–538. https://doi.org/10.1016/j.autcon.2009.11.004.

Gu, N. and London, K. (2010). Understanding and facilitating BIM adoption in the AEC industry. *Automation in Construction*, 19, 988–999. https://doi.org/10.1016/j.autcon.2010.09.002.

Harding, J. (2012). Avoiding project failures. *Chemical Engineering*, 119(12), 51–54. https://search-proquest-com.libproxy1.nus.edu.sg/docview/1248290016?pq-origsite=summon.

Hewage, K. N., Ruwanpura, J. Y., and Jergeas, G. F. (2008). IT usage in Alberta's building construction projects: Current status and challenges. *Automation in Construction*, 17(8), 940–947. https://doi.org/10.1016/j.autcon.2008.03.002

Hwang, B.-G., Ngo, J., and Her, P. W. Y. (2020). Integrated digital delivery: Implementation status and project performance in the Singapore construction industry. *Journal of Cleaner Production*, 262, 121396. https://doi.org/10.1016/j.jclepro.2020.121396.

Hwang, B.-G. and Tan, J. S. (2012). Green building project management: Obstacles and solutions for sustainable development. *Sustainable Development*, 20(5), 335–349. https://doi.org/10.1002/sd.492.

Hwang, B. G. (2018). Performance and improvement of green construction projects. In *Performance and Improvement of Green Construction Projects*. https://doi.org/10.1016/c2017-0-01403-9.

Jia, M., Komeily, A., Wang, Y., and Srinivasan, R. S. (2019). Adopting Internet of Things for the development of smart buildings: A review of enabling technologies and applications. *Automation in Construction*, 101, 111–126. https://doi.org/10.1016/j.autcon.2019.01.023.

Jiang, L., Li, Z., Li, L., and Gao, Y. (2018). Constraints on the promotion of prefabricated construction in China. *Sustainability*, 10(7). https://doi.org/10.3390/su10072516.

Kunz, J. and Fischer, M. (2012). Virtual design and construction: Themes, case studies and implementation suggestions. https://stacks.stanford.edu/file/druid:gg301vb3551/WP097_0.pdf.

Leviäkangas, P., Paik, S. M., and Moon, S. (2017). Keeping up with the pace of digitization: The case of the Australian construction industry. *Technology in Society*, 50, 33–43. https://doi.org/10.1016/j.techsoc.2017.04.003.

Liu, D., Lu, W., and Niu, Y. (2018). Extended technology-acceptance model to make smart construction systems successful. *Journal of Construction Engineering and Management*, 144(6), 04018035. https://doi.org/10.1061/(ASCE)CO.1943-7862.0001487.

Loosemore, M. (2014). Improving construction productivity: A subcontractor's perspective. In *Engineering, Construction and Architectural Management*, 21(3), 245–260). https://doi.org/10.1108/ECAM-05-2013-0043.

Lunn, S. D. and Stephenson, P. (2000). The impact of tactical and strategic FM automation. *Facilities*, 18(7–8), 312–323. https://doi.org/10.1108/02632770010340708.

Mitropoulos, P. and Tatum, C. B. (2002). Forces driving adoption of new information technologies. *Journal of Construction Engineering and Management*, 126(5), 340–348. https://doi.org/10.1061/(asce)0733-9364(2000)126:5(340).

Nagy, J., Oláh, J., Erdei, E., Máté, D., and Popp, J. (2018). The role and impact of Industry 4.0 and the Internet of Things on the business strategy of the value Chain — the case of Hungary. *Sustainability*, 10(10), 1–25. https://doi.org/10.3390/su10103491.

Ng, J. Sen. (2017, October 24). 80,000 Singaporeans to be trained in construction technology by 2025. The Straits Times. https://www.straitstimes.com/singapore/housing/80000-singaporeans-to-be-trained-in-construction-technology-by-2025.

Ngo, J., Hwang, B.-G., and Her, P. W. Y. (2020). Integrated digital delivery: Implementation status, barriers, and strategies. In *Construction Research Congress 2020*, pp. 733–742, Tempe, Arizona: ASCE. https://doi.org/10.1061/9780784482865.078.

Oesterreich, T. D. and Teuteberg, F. (2016). Understanding the implications of digitisation and automation in the context of Industry 4.0: A triangulation approach and elements of a research agenda for the construction industry. *Computers in Industry*, 83, 121–139. https://doi.org/10.1016/j.compind.2016.09.006.

Oesterreich, T. D. and Teuteberg, F. (2019). Behind the scenes: Understanding the socio-technical barriers to BIM adoption through the theoretical lens of information systems research. *Technological Forecasting and Social Change*, 146, 413–431. https://doi.org/10.1016/j.techfore.2019.01.003.

Rashid, S. (2018, November 14). Construction sector gets S$4 million to boost digitalization. Channel NewsAsia. https://www.channelnewsasia.com/news/business/bca-mnd-construction-digitalisation-data-integration-10929154.

Rose, T. M. and Manley, K. (2012). Adoption of innovative products on Australian road infrastructure projects. *Construction Management and Economics*, 30, 277–298. https://doi.org/10.1080/01446193.2012.665173.

Sepasgozar, S. M. E., Loosemore, M., and Davis, S. R. (2016). Conceptualizing information and equipment technology adoption in construction: A critical review of existing research. *Architectural Management*, 23, 158–176. https://doi.org/10.1108/ECAM-05-2015-0083.

Smith, P. (2014). BIM implementation — global strategies. *Procedia Engineering*, 85, 482–492. https://doi.org/10.1016/j.proeng.2014.10.575.

Stewart, R. A., Mohamed, S., and Marosszeky, M. (2004). An empirical investigation into the link between information technology implementation barriers and coping strategies in the Australian construction industry. *Construction Innovation*, 4(3), 155–171. https://doi.org/10.1108/147141704 10815079.

Suprun, E. V. and Stewart, R. A. (2015). Construction innovation diffusion in the Russian Federation. *Construction Innovation*, 15(3), 278–312. https://doi.org/10.1108/ci-07-2014-0038.

Tatum, C. B. (1987). Process of innovation in construction firm. *Journal of Construction Engineering and Management*, 113(4), 648–663. https://doi.org/10.1061/(ASCE)0733-9364(1987)113:4(648).

Venkatesh, V. and Bala, H. (2008). Technology acceptance Model 3 and a research agenda on interventions. *Decision Sciences*, 39(2), 273–315. https://doi.org/10.1111/j.1540-5915.2008.00192.x.

Wong, S. Y. and Ng, J. Sen. (2017, June 19). Construction industry faces tough times. The Straits Times. https://www.straitstimes.com/singapore/construction-industry-faces-tough-times.

World Economic Forum. (2018). Shaping the future of construction future scenarios and implications for the industry.

Yik, F. W. H., Lai, J. H. K., Chan, K. T., and Yiu, E. C. Y. (2006). Problems with specialist subcontracting in the construction industry. *Building Services Engineering Research and Technology*, 27(3), 183–193. https://doi.org/10.1191/0143624406bse160oa.

## Chapter 3

American Institute of Architects. (2008). A history of design-build. Web Presentation. Downloaded April 30, 2021. https://network.aia.org/DownloadDocumentFile.pdfild.

Anonymous. (2020 July). Interview with official of the state's DOT. The official wishes to remain anonymous.

Bradbury, O. (2018). The overlooked influence of John Soane on architecture from 1791 to 1980 critical appraisal. A thesis submitted in partial fulfilment of the requirements of Sheffield Hallam University for the degree of Doctor of Philosophy.

Design-Build Institute of America. What is design-build? Retrieved July 7, 2020. https://dbia.org/what-is-design-build/.

Federal Highway Administration. (2016). Defining PPPs and their benefits. Retrieved August 6, 2020. https://www.fhwa.dot.gov/reports/pppwave/03.htm.

Federal Highway Administration. (2018). Alternative contracting method performance in U.S. highway construction. FHWA Publication No: FHWA-HRT-17-100. A technical summary of the Federal Highway Administration report "Alternative Contracting Method Performance in U.S. Highway Construction" (DTFH61-13-C-00024).

Interview with Doug Gransberg at the meeting of the Transportation Research Board, January 2017.

Gross, M. E. and Garvin, M. J. (2011). Structuring PPP toll-road contracts to achieve public objectives. *Engineering Project Organization Journal*, 1(2), 143–156.

Ibbs, C., Kwak, Y., Ng, T., and Odabasi, A. (2003). Project delivery systems and project change: Quantitative analysis. *Journal of Construction Engineering and Management*, 129(4), 382–387.

King James Bible, I Cor. 3:10. First published 1611.

Leiringer, R. (2006). Technological innovation in PPPs: Incentives, opportunities and actions. *Journal of Construction Management & Economics*, 24(3), 301–308.

Mahmudova, M. (2013). PPP/PFI Outlook FY 2012. *Infrastructure Journal*, 4, 4 April 2013.

Minchin, R. E. and Gross, M. E. (2015). Project delivery systems and innovation: The case of US road building. Construction Innovation, Chapter 12, Oxford, UK: Blackwell Publishing, Ltd. WILEY, pp. 165–180.

Minchin, R. E., Migliaccio, G. C., Atkins, K., Warne, T., Gatti, U. (g), Ptschelinzew, L.(g.), Hostetler, G., and Asiamah, S. (2014). *Design Management Guide for Design-Build and Construction-Manager-as-General-Contractor Projects*. Washington, DC: Transportation Research Board of the National Academies.

Minchin, R. E., Tran, D., Gad, G., Scott, S., Konrath, L., D'Angelo, D., Russell, J., Schexnayder, C., Myers, M., and Smadi, A. (2019). Framework for implementing constructability across the entire project development process: NEPA to final design. NCHRP 10-99 Interim Report, Washington, DC.

US Department of Transportation. (2009). Public-private partnerships for highway infrastructure: Capitalizing on international experience. Report No. FHWA-PL-09-010, Washington, DC.

Yale Law School. The code of Hammurabi. Retrieved July 6, 2020. https://avalon.law.yale.edu/ancient/hamframe.asp.

## Chapter 4

Carpenter, J. (2012). Annual report on alternate technical concept programmatic Waiver, SEP-14 Progress Report, Washington State DOT, Olympia, January 31, p. 6.

Chasey, A. and Schexnayder, A. (2000). Constructability: The key to reducing investment risk. *Acta Structilia*, 7(1&2), 93–112.

Chong, H., Preece, C., and Rogers, J. (2014). BIM update 2013: A mixed review approach from academia and industry. *Trends and Development in Management Studies*, 3(1), 1–21.

Coblentz, M. A. (2012). Alternative technical concepts used on the ICC project. 91st Annual Meeting of the Transportation Research Board, Paper #12-6806, Washington, D.C., p. 17.

Federal Highway Administration (FHWA). (2012). Alternative technical concepts. Unpublished presentation, Every Day Counts 2, U.S.DOT, FHWA, Washington, D.C. http://www.fhwa.dot.gov/accelerating/presentations/atc/.

Ford, D. N., Anderson, S. D., Damron, A. J., de Las Casas, R., Gokmen, N., and Kuennen, S. T. (2004). Managing constructability reviews to reduce highway project durations. *Journal of Construction Engineering and Management*, 130(1), 33–42.

Gad, G. M., Gransberg, D. D., and Loulakis, M. (2015). Policies and procedures for successful implementation of alternative technical concepts. *Transportation Research Record*, 2504(1), 78–86.

Gad, G. M., Loulakis, M. C., and Gransberg, D. (2020). Confidentiality issues with alternative technical concepts during design-build project procurement. *Journal of Legal Affairs and Dispute Resolution in Engineering and Construction*, 12(1), 04519034.

Gransberg, D., Loulakis, M. C., and Gad, G. M. (2014). Alternative technical concepts for contract delivery methods (Project No. 20-05 -Topic 44-09). National Commission of Highway Research.

Hiley, A. and Yagci, O. (2001). The implementation of constructability: A prerequisite in raising the quality of project outcome. In Akintoye, A. (Ed.), *17th Annual ARCOM Conference*, 5–7 September 2001, University of Salford. Association of Researchers in Construction Management, Vol. 1, pp. 261–270.

Jergeas, G. and Put, J. (2001). Benefits of constructability on construction projects. *Journal of Construction Engineering and Management*, 127(4), 281–290.

Jiang, L. and Leicht, R. M. (2014). Automated rule-based constructability checking: Case study of formwork. *Journal of Management in Engineering*, 31(1), A4014004.

Khan, S. (2015). An overview of constructability: A management tool for architects. *Architecture Research*, 5(5), 125–139.

Kuo, V. and Wium, J. A. (2014). The management of constructability knowledge in the building industry through lessons learnt programmes. *Journal of the South African Institution of Civil Engineering*, 56(1), 20–27.

Mao, X., Zhang, X., and AbouRizk, S. M. (2009). Enhancing value engineering process by incorporating inventive problem-solving techniques. *Journal of Construction Engineering and Management*, 135(5), 416–424.

Molenaar, K. R. and Gransberg, D. D. (2001). Design-builder selection for small highway projects. *Journal of Management in Engineering*, 17(4), 214–223.

National Cooperative Highway Research Program. (2019). Framework for implementing constructability across the entire project development process: NEPA to final design. Interim Report (not published).

O'Connor, J. T. and Tucker, R. L. (1986). Industrial project constructability improvement. *Journal of Construction Engineering and Management*, 112(1), 69–82.

Russel, J. S., Swiggum, K. E., Shapiro, J. M., and Alaydrus, A. F. (1994). Constructability related to TQM, value engineering, and cost/benefits. *Journal of Performance of Constructed Facilities*, 8(1), 31–45.

Utah Department of Transportation. (2019). UDOT project preview information. Accessed May 15, 2021. https://sites.google.com/utah.gov/udot-project-previews/home.

Utah Department of Transportation. (2019). SR-7 Southern parkway: Extend. Accessed May 15, 2021. https://sites.google.com/a/utah.gov/southernparkway/industry-rfi-and-comments.

## Chapter 5

Ahmad, T., Aibinu, A. A., Stephan, A., and Chan, A. P. (2019). Investigating associations among performance criteria in Green Building projects. *Journal of Cleaner Production*, 232, 1348–1370.

Chan, Y. and Walmsley, R. P. (1997). Learning and understanding the Kruskal-Wallis one-way analysis-of-variance-by-ranks test for differences among three or more independent groups. *Physical Therapy*, 77(12), 1755–1761.

Ding, Z., Fan, Z., Tam, V. W., Bian, Y., Li, S., Illankoon, I. C. S., and Moon, S. (2018). Green building evaluation system implementation. *Building and Environment*, 133, 32–40.

Dunn, O. J. (1964). Multiple comparisons using rank sums. *Technometrics*, 6(3), 241–252.

Eichholtz, P., Kok, N., and Quigley, J. M. (2013). The economics of green building. *Review of Economics and Statistics*, 95(1), 50–63.

Ferme, L., Zuo, J., and Rameezdeen, R. (2018). Improving collaboration among stakeholders in green building projects: Role of early contractor involvement. *Journal of Legal Affairs and Dispute Resolution in Engineering and Construction*, 10(4), 04518020.

Frigge, M., Hoaglin, D. C., and Iglewicz, B. (1989). Some implementations of the boxplot. *The American Statistician*, 43(1), 50–54.

Green Building Council of Australia, G. (2020). Green Star project directory. Accessed January 12, 2019. https://www.gbca.org.au/project-directory.asp.

Guan, L., Abbasi, A., and Ryan, M. J. (2020). Analyzing green building project risk interdependencies using Interpretive Structural Modeling. *Journal of Cleaner Production*, 256, 120372.

Gunhan, S. (2019). Analyzing sustainable building construction project delivery practices: Builders' perspective. *Practice Periodical on Structural Design and Construction*, 24(1), 05018003.

Gunhan, S. and Hatipkarasulu, Y. (2012). Scope of preconstruction services in green building projects. Paper presented at the American Society for Engineering Education.

Hwang, B.-G., Zhao, X., and Tan, L. L. (2015). Green Building Projects: Schedule Performance, influential factors and solutions. *Engineering, Construction and Architectural Management*, 22(3), 327–346. https://doi.org/10.1108/ecam-07-2014-0095

Hwang, B.-G., Zhu, L., Wang, Y., and Cheong, X. (2017). Green building construction projects in Singapore: Cost premiums and cost performance. *Project Management Journal*, 48(4), 67–79.

Ibrahim, M. W., Hanna, A., and Kievet, D. (2020). Quantitative comparison of project performance between project delivery systems. *Journal of Management in Engineering*, 36(6), 04020082.

IEA. (2019). 2019 Global status report for buildings and construction. United Nations Environment Programme.

Iheanyichukwu Joachim, O., Kamarudin, N., Uche Aliagha, G., and Ufere, K. J. (2015). Theoretical explanations of environmental motivations and expectations of clients on green building demand and investment. *E&ES*, 23(1), 012010.

Kantola, M., and Saari, A. (2016). Project Delivery Systems for NZEB projects. *Facilities*, 34(1/2), 85–100. https://doi.org/10.1108/f-03-2014-0025

Miller, S. A., Horvath, A., and Monteiro, P. J. (2018). Impacts of booming concrete production on water resources worldwide. *Nature Sustainability*, 1(1), 69–76.

Molenaar, K. and Franz, B. (2018). *Revisiting Project Delivery Performance*. Charles Pankow Foundation and Construction Industry Institute. Austin, Texas: Construction Industry Institute.

Molenaar, K. R., Sobin, N., and Antillón, E. I. (2010). A synthesis of best-value procurement practices for sustainable design-build projects in the public sector. *Journal of Green Building*, 5(4), 148–157.

OECD. (2019). *Global Material Resources Outlook to 2060 Economic Drivers and Environmental Consequences*. OECD Publishing.

Olanipekun, A. O., Chan, A. P., Xia, B. P., and Ameyaw, E. E. (2017). Indicators of owner commitment for successful delivery of green building projects. *Ecological Indicators*, 72, 268–277.

Olanipekun, A. O., Xia, B., Hon, C., and Darko, A. (2018). Effect of motivation and owner commitment on the delivery performance of green building projects. *Journal of Management in Engineering*, 34(1), 04017039.

Olanipekun, A. O., Xia, B., Hon, C., and Hu, Y. (2017). Project owners' motivation for delivering green building projects. *Journal of Construction Engineering and Management*, 143(9), 04017068.

Orsi, A., Guillén-Guillamón, I., and Pellicer, E. (2020). Optimization of green building design processes: Case studies within the European Union. *Sustainability*, 12(6), 2276.

Orsi, A., Pellicer, E., and Guillamon, I. G. (2017). Optimizing Green-Building Project Delivery: Comparative Analysis of Design-bid-build and design-build delivery methods using case studies. *Proceedings of International Structural Engineering and Construction*, 4(1). https://doi.org/10.14455/isec.res.2017.202.

Raouf, A. M. and Al-Ghamdi, S. G. (2019a). Effectiveness of project delivery systems in executing green buildings. *Journal of Construction Engineering and Management*, 145(10), 0311005.

Raouf, A. M. and Al-Ghamdi, S. G. (2019b). Framework to optimize cost of quality in delivering and operating green buildings. Paper presented at the International Conference on Sustainable Infrastructure 2019: Leading Resilient Communities through the 21st Century.

Shapiro, J. (2008). Percentage frequency distribution. *Encyclopedia of Survey Research Methods*, 578–579.

Tang, Z., Ng, S. T., and Skitmore, M. (2019). Influence of procurement systems to the success of sustainable buildings. *Journal of Cleaner Production*, 218, 1007–1030.

West, N. and Gransberg, D. (2012). Quantifying the value of construction contractor preconstruction involvement. Paper presented at the First Australasia and South East Asia Structural Engineering and Construction Conference (ASEA-SEC-1) Perth, Australia.

Zhang, J., Li, H., Olanipekun, A. O., and Bai, L. (2019). A successful delivery process of green buildings: The project owners' view, motivation and commitment. *Renewable Energy*, 138, 651–658.

## Chapter 7

Adamtey, S. A. (2020). An empirical study to assist owners in selecting the right procurement method for design-build projects. *International Journal of Construction Education and Research,* 18, 1–18.

Alleman, D., Antoine, A., Gransberg, D. D., and Molenaar, K. R. (2017). Comparison of qualifications-based selection and best-value procurement for construction manager–general contractor highway construction. *Transportation Research Record*, 2630(1), 59–67.

Asmar, E. M. (2012). Modeling and benchmarking performance for the integrated project delivery (Ipd) system (dissertation). University of Wisconsin--Madison, Madison, WI.

Gransberg, D. D. and Ellicott, M. A. (1997). Best-value contracting criteria. *Cost Engineering*, 39(6), 31. AACE International.

Gransberg, D. and Touran, A. (2019). The value, benefits, and limitations of qualifications-based selection (QBS) for airport project delivery. ACRP Synthesis 102. Airport Cooperative Research Program.

Molenaar, K. and Franz, B. (2018). Revisiting project delivery performance. Charles Pankow Foundation and Construction Industry Institute.

Ramsey, D., El Asmar, M., and Gibson Jr, G. E. (2016). Quantitative performance assessment of single-step versus two-step design-build procurement. *Journal of Construction Engineering and Management*, 142(9), 04016033. American Society of Civil Engineers.

Shelton, B. (2018). *Qualifications-Based Selection (QBS): Best Practice for Architecture, Engineering and Construction Management/General Contractor Procurement in Canada*. ON, Canada: A QBS Canada Publication.

## Chapter 8

Abdelrahman, M., Zayed, T., and Elyamany, A. (2008). Best-value model based on project specific characteristics. *Journal of Construction Engineering and Management*, 134(3), 179–188. https://doi.org/10.1061/(ASCE)0733-9364(2008)134:3(179).

Anderson, S. D. and Russell, J. S. (2001). Guidelines for warranty, multi-parameter, and best value contracting. In NCHRP Report (Vol. 451).

Asmar, E. M., Ramsey, D., and Gibson, G. E. (2020). Design-build for transportation projects: Cost and schedule change performance analysis. *Journal of Legal Affairs and Dispute Resolution in Engineering and Construction*, 12(1), 04519030.

Calahorra-Jimenez, M., Molenaar, K. R., Torres-Machi, C., Chamorro, A., and Alarcón, L. F. (2020). Structured approach for best-value evaluation criteria: US design-build highway procurement. *Journal of Management in Engineering*, 36(6), 1–14. https://doi.org/10.1061/(ASCE)ME.1943-5479.0000857.

Calahorra-Jimenez, M., Torres-Machi, C., Chamorro, A., Alarcon, L. F., and Molenaar, K. R. (2021). The importance of non-cost criteria weighting in best-value design-build U.S. highway projects. *Journal of Management in Engineering*, 37(4), 04021027. https://doi.org/10.1061/(ASCE)ME.1943-5479.0000912.

D'Agostino, R. B. and Stephens, M. A. (1986). *Goodness-of-Fit Techniques*. (Vol. 151, Issue 1). New York: Marcel Dekker.

Ellis, R., Herbsman, Z., and Kumar, A. (1991). DBProgramEval1991.pdf.

FHWA. (2006). Study to Congress on the Effectiveness of Design-Build Project Delivery Relating to the Federal-Aid Highway Program.

FHWA. (2018). Alternative Contracting Method Performance in US Highway Construction. Techbrief.

FMI (Fails Management Institute). (2018). Design-build utilization. Combined market study.

Franz, B., Molenaar, K. R., and Roberts, B. A. M. (2020). Revisiting project delivery system performance from 1998 to 2018. *Journal of Construction Engineering and Management*, 146(9), 04020100.

Gaikwad, S. V., Calahorra-Jimenez, M., Molenaar, K. R., and Torres-Machi, C. (2021). Challenges in engineering estimates for best-value design-build projects: An analysis of bid dispersion in U.S. highway projects. *Journal of Construction Engineering and Management*, 147(7), 04021065. https://doi.org/DOI: 10.1061/(ASCE)CO.1943-7862.0002104.

Gransberg, D. D., Koch, J. E., and Molennar, K. R. (2006). Design-build: What and why. In *Preparing for Design-Build Projects. A Primer for Owners, Engineers, and Contractors*. American Society of Engineers (ASCE). p. 7.

Gumbel, E. J. (1955). Statistical theory of extreme values and some practical applications. In *The Mathematical Gazette* (Vol. 39, Issue 330). https://doi.org/10.2307/3608620.

Hale, D. R., Shrestha, P. P., Gibson, G. E., and Migliaccio, G. C. (2009). Empirical comparison of design/build and design/bid/build project delivery methods. *Journal of Construction Engineering and Management*, 135(7), 579–587.

Molenaar, K. R., Gransberg, D., Scott, S., Down, D., and Ellis, R. (2005). Recommended AASHTO Design — Build Procurement Guide. 20.

Molenaar, K. R. and Tran, D. (2015). NCHRP Synthesis 471. Practices for Developing Transparent Best Value Selection Procedures. Transportation Research Board of the National Academies. https://doi.org/10.17226/22192.

Molenaar, K. R., Harper, C., and Yugar-Arias, I. (2014). Guidebook for selecting alternative contracting methods for roadway projects: Project delivery methods, procurement procedures, and payment provisions.

Mun, J. (2002). *Advanced Analytical Models*. John Wiley & Sons.

Mun, J. and Pomering, R. (2010). *Applying Monte Carlo Risk Simulation, Strategic Real Options, Stochastic Forecasting, and Portfolio Optimization*. New York: John Wiley & Sons.

Pakkala, P. (2002). Innovative Project Delivery Methods for Infrastructure. Finnish Road Enterprise, Helsinki, 19.

Riley, D. R., Diller, B. E., and Kerr, D. (2005). Effects of delivery systems on change order size and frequency in mechanical construction. *Journal of Construction Engineering and Management*, 131(9), 953–962.

Scott, S., Molenaar, K. R., Gransberg, D. D., and Smith, N. C. (2006). NCHRP Report 561. Best-value procurement methods for highway construction projects. Transportation Research Board of the National Academies. https://doi.org/10.17226/13982.

Shonkwiler, R. W. and Mendivil, F. (2009). Introduction to Monte Carlo Methods. In *Explorations in Monte Carlo Methods*. Springer. https://doi-org.hmlproxy.lib.csufresno.edu/10.1007/978-0-387-87837-9_1.

Shrestha, P. P., O'Connor, J. T., and Gibson, G. E. (2012). Performance comparison of large design-build and design-bid-build highway projects. *Journal of Construction Engineering and Management*, 138(1), 1–13.

Shrestha, P. P. and Fernane, J. D. (2017). Performance of design-build and design-bid-build projects for public universities. *Journal of Construction Engineering and Management*, 143(3), 04016101.

The American Institute of Architects. (2007). Integrated project delivery: A guide. In *The American Institute of Architects* (Vol. 1, Issue 1). https://doi.org/10.1016/j.autcon.2010.09.002.

Tran, D. Q., Diraviam, G., and Minchin, R. E. (2018). Performance of highway design-bid-build and design-build projects by work types. *Journal of Construction Engineering and Management*, 144(2), 04017112.

Tran, Dai Q, Molenaar, K. R., and Kolli, B. (2017). Implementation of best-value procurement for highway design and construction in the USA. *Engineering, Construction and Architectural Management*, 24(5), 774–787.

Code of Federal Regulations. Part 636. DB Contracting, 212 (2002).

Yoe, C. (2019). Choosing a probability distribution. In *Principles of Risk Analysis. Decision Making Under Uncertainty*. Taylor & Francis Group, LLC, pp. 319–361. https://doi.org/10.1201/b11256-12.

## Chapter 9

EN 15804 — 2012 Sustainability of construction works, environmental product declarations, core rules for the product category of construction products. U.S. Green Building Council. 2018. Usgbc.Org. https://www.usgbc.org/resources/en-15804%E2%80%942012-sustainability-construction-works-environmental-product-declarations-core-ru.

ISO 14025:2006 — Environmental labels and declarations — type III environmental declarations — principles and procedures. 2018. Iso.Org. https://www.iso.org/standard/38131.html.

ISO 14040:2006 — Environmental management — life cycle assessment — principles and framework. 2018. Iso.Org. https://www.iso.org/standard/37456.html.

ISO 14044:2006 — Environmental management — life cycle assessment — Requirements and Guidelines. 2018. Iso.Org. https://www.iso.org/standard/38498.html.

ISO 21930:2017 — Sustainability in buildings and civil engineering works — core rules for environmental product declarations of construction products and services. 2018. Iso.Org. https://www.iso.org/standard/61694.html.

Al-Qadi, I. L., Yang, R., Kang, S., Ozer, H., Ferrebee, E., Roesler, J. R., Salinas, A., Meijer, J., Vavrik, W. R., and Gillen, S. L. (2015). Scenarios developed for improved sustainability of Illinois Tollway. *Transportation Research Record: Journal of the Transportation Research Board,* 2523, 11–18. https://doi.org/10.3141/2523-02.

Aurangzeb, Q., Al-Qadi, I. L., Ozer, H., and Yang, R. (2014). Hybrid life cycle assessment for Asphalt mixtures with high RAP content. *Resources, Conservation and Recycling*, 83, 77–86. https://doi.org/10.1016/j.resconrec.2013.12.004.

Bhat, C. G. and Mukherjee, A. (2019). Sensitivity of life-cycle assessment outcomes to parameter uncertainty: Implications for material procurement decision-making. *Transportation Research Record: Journal of the Transportation Research Board*, 2673(3), 106–114.

Bhat, C. G. and Mukherjee, A. (2020). A life-cycle thinking informed approach to support pavement design decision-making. *ASCE Journal of Transportation Engineering: Part B, Pavements*, 146(4), 04020067.

Buy Clean California Act. (2017). Vol. -262, 816. California: Legislative Counsel Bureau, State of California.

European Union (EU). (2021). EU GPP criteria. Website. Accessed 20 June 2021. https://ec.europa.eu/environment/gpp/eu_gpp_criteria_en.htm.

Ferraro, F., Etzion, D., and Gehman, J. (2015). Tackling grand challenges pragmatically: Robust action revisited. *Organization Studies*, 36(3), 363–390. https://doi.org/10.1177/0170840614563742.

Korkmaz, S., Riley, D., and Horman, M. (2010). Piloting evaluation metrics for sustainable high-performance building project delivery. *Journal of Construction Engineering and Management*, 136(8), 877–885.

Muench, S.T., Migliaccio, G., Mukherjee, A., Anderson, J., Kaminsky, J., Ashtiani, M. Z., and Bhat, C. (2019). A guidebook for sustainable highway construction practices. TRB's National Cooperative Highway Research Program (NCHRP) Research Report: 916 (10-91A).

Mukherjee, A. and Miller, L. (2019). Minnesota DOT EPD Pilot Project: Development of a Sustainable Procurement Roadmap. Minnesota Department of Transportation, Research Report, MN/RC 2019-39.

Mukherjee, A., Bhat, C., and Harvey, J. (2020). Challenges in meeting data needs for use of environmental product declarations in pavement design and construction: State of practice and future scope. FHWA-HRT- 20-022; Federal Highway Administration.

Mukherjee, A. and Dylla, H. (2017). Challenges to using environmental product declarations in communicating life-cycle assessment results. *Transportation Research Record: Journal of the Transportation Research Board,* 2639, 84–92. https://doi.org/10.3141/2639-11.

OECD. (2015). Going Green: Best practices for sustainable procurement. Organization for Economic Co-operation and Development, Paris, France.

Scott, S., Molenaar, K. R., Gransberg, D. D., and Smith, N. C. (2006). NCHRP Report 561: Best-value procurement methods for highway construction projects. Transportation Research Board of the National Academies, Washington, D.C.

Simonen, K. (2019). Buy clean Washington study. https://carbonleadershipforum.org/buy-clean-washington-study/.

United Nations Economic and Social Council. (2020). Progress report on the 10-year framework of programmes on sustainable consumption and production patterns. 2020 session, 25 July 2019–20 July 2020, agenda item 6.

U.S. General Services Administration Federal Acquisition Services. (2020). Federal Procurement Data System Product and Service Codes (PSC) Manual. Fiscal Year 2021 Edition. https://www.acquisition.gov/sites/default/files/manual/PSC_Manual_October_2020.pdf.

Van Dam, T. J., Harvey, J. T., Muench, S. T., Smith, K. D., Snyder, M. B., Al-Qadi, I. L., and Ozer, H., *et al.* (2015). Towards sustainable pavement systems: A reference document. FHWA-HIF-15-002; Federal Highway Administration.

Zietsman, J., Ramani, T., Potter, J., Reeder, V., and DeFlorio, J. (2011). NCHRP Report 708: A guidebook for sustainability performance measurement for transportation agencies. Transportation Research Board, National Academies, Washington, D.C.

# Chapter 11

AGC of America (AGC). (2009). Qualifications based selection of contractors. https://www.agc.org/sites/default/files/Qualifications%20Based%20 Selection%20of%20Contractors%20FINAL.PDF.

Asian Development Bank (ADB). (2018). Guide to performance-based road maintenance contracts.

Frost, M. (2001). Imperatives in future road system management, improved road maintenance productivity, the Australian case. 14th IRF Road World Congress, Paris, France.

Frost, M. and Lithgow, C. (1996). Improving quality and cutting cost through performance contracts — Australian experience. Paper prepared for Road Management Training Seminar of the World Bank, December 17–18. http://www.zietlow.com/docs/frost.htm.

Gordon, S. B. (1998). Performance Incentive Contracting: Using the Purchasing Process to Find Money Rather Than Spend It. GOVERNMENT FINANCE REVIEW, 14, 33–37.

Gericke, B., Henning, T., and Greewood, I. (2014). Review of performance based contracting in the road sector — Phase 1. The World Bank.

Hartwig, T., Mumssen, Y., and Schliessler, A. (2005). Output-based aid in Chad: Using performance-based contracts to improve roads. OB Approaches 33160.

Hinze, J. (2011). *Construction Contracts*. 3rd Ed., US: McGraw-Hill. ISBN: 978-0-07-339785-6.

Iowa DOT. (2015). Standard specifications for highway and bridge construction.

Legal Information Institute (LII). (2021). 37.101 Definitions. https://www.law.cornell.edu/cfr/text/48/37.101.

McCullouch, B. G., Sinha, K. C., and Anastasopoulos, P. C. (2009). Performance-based contracting for roadway maintenance operations in Indiana. Final Report, FHWA/IN/JTRP-2008/12.

NCHRP. (2009). Performance-based contracting for maintenance. Washington, D.C.: National Cooperative Highway Research Program (NCHRP Synthesis 389). http://onlinepubs.trb.org/onlinepubs/nchrp/nchrp_syn_389.pdf.

Organization for Economic Co-operation and Development (OECD). (2006). Guidelines for performance-based contracts between municipalities and water utilities in Eastern Europe, Caucasus and Central Asia (EECCA). https://www.oecd.org/environment/outreach/36397942.pdf.

Pakkala, P. (2002). Innovative project delivery methods for infrastructure, an international perspective. Finnish Road Enterprise.

Pakkala, P. (2005). Performance-based contracts — International experiences. Finnish Road Administration, TRB Executive Workshop — April.

Popescu, L. and Monismith, C. L. (2006). Performance-based pay factors for Asphalt concrete construction: Comparison with a currently used experience-based approach. Research Report: UCPRC-RR-2006-16.

Ribreau, N. (2003). Synopsis of WSDOT's review of highway maintenance "Outsourcing" experience. Transportation Research Board Committee A3C01 Maintenance and Operations Management.

Schexnayder, C. and Ohrn, L. G. (1997). Highway specifications — quality versus pay. *Journal of Construction Engineering and Management*, ASCE, 123(4), 437–443.

Shrestha, K. (2016). Framework of performance-based contracting for chip seal and striping maintenance activities. Ph.D. Dissertation, UNLV.

Shrestha, K., Deshar, R., Kisi, K. P., and Taffazzoli, T. (2021). Data digitization to mitigate legal disputes during reconstruction in earthquake prone areas. *Journal of Legal Affairs and Dispute Resolution in Engineering and Construction*, 13(2), 05021001, ASCE.

Shrestha, K. and Powers, M. (2018). Performance-based contracting for maintaining rest area stops. 54th ASC Annual International Conference Proceedings, Associated Schools of Construction.

Shrestha, K. and Shrestha, P. P. (2016). Cost and customer satisfaction of culvert cleaning and sweeping road projects in Nevada. Proceeding of Construction Research Congress, ASCE, San Juan, Puerto Rico, May 31–June 2.

Shrestha, P. P. and Shrestha, K. (2014a). An evaluation of performance based maintenance contracting methods. Proceeding of Construction Research Congress, ASCE, Atlanta, May 19–21.

Shrestha, K. and Shrestha, P. P. (2014b). A tool to select road maintenance contracting methods. Proceeding of Construction Research Congress, ASCE, Atlanta, May 19–21.

Shrestha, K. and Shrestha, P. P. (2013). Framework development for cost comparison of DOT contracting methods. Canadian Society for Civil Engineering Annual Conference, CSCE, Montreal, Canada, May 29-June 1.

Shrestha, K., Shrestha, P. P., and Lidder, M. (2019). Life-cycle cost comparison of chip seal and striping: In-house workers versus private contractors. *Journal of Engineering, Construction and Architectural Management, Emerald Insight*, 26(6), 927–944.

Shrestha, K. and Shrestha, P. P. (2020). Framework development of performance-based striping maintenance contracts. *Journal of Legal Affairs and Dispute Resolution in Engineering and Construction*, 12(2), ASCE.

Shrestha, P. P., Shrestha, K., and Lidder, M. (2017). Quality of road maintenance work performed by state force versus private contractors in Nevada. *Journal of Practice Periodical on Structural Design and Construction,* 22(4), ASCE.

Shrestha, P. P., Shrestha, K., Said, A., and Lidder, M. (2016). Factors affecting the selection of in-house and outsourcing road maintenance methods and assessment of their benefits. *Journal of Infrastructure Systems,* 23(1), ASCE.

Southern Area Consortium of Human Services (SACHS). (2004). Innovations in performance-based contracting. https://www.cdss.ca.gov/ccr/res/pdf/performance/1.%203InnovationPBCReport9-14-04.pdf.

Stankevich, N., Qureshi, N., and Querioz, C. (2005). Performance-based contracting for preservation and improvement of road assets. Transport Note No. TN-27, World Bank, Washington D.C.

Stankevich, N., Qureshi, N., and Querioz, C. (2006). Resource guide, performance-based contracting for preservation and improvement of road assets. World Bank, Washington D.C.

Stankevich, N., Qureshi, N., and Queiroz, C. (2009). Performance-based contracting for preservation and improvement of road assets. Transport Notes 33947, The World Bank, Washington D.C.

Straub, A. (2009). Cost savings from performance-based maintenance contracting. *International Journal of Strategic Property Management*, 13(3), 205–217. https://doi.org/10.3846/1648-715X.2009.13.205-217.

Wirahadikusumah, R., Susanti, B., Coffey, V., and Adighibe, C. (2015). Performance-based contracting for roads — experiences of Australia and Indonesia. *Procedia Engineering*, 125, 5–11.

USDOT, FHWA. (2013). Performance contracting improves project outcomes. https://www.youtube.com/watch?v=m9rJ7WLzcO4.

University of Reading (UR). (2013). Performance-based contracting in the construction sector. School of Construction Management and Engineering. https://sci-network.eu/fileadmin/templates/sci-network/files/Resource_Centre/Reports/Performance-based_contracting_in_the_construction_sector.pdf.

Zietlow, G. (2004). Implementing performance-based road management and maintenance contracts in developing countries — an instrument of German technical cooperation. German Development Cooperation (GTZ).

Zietlow, G. (2008). Cutting costs and improving quality through performance-based road management and maintenance contracts — the Latin American and OECD experiences. https://ppiaf.org/sites/ppiaf.org/files/documents/

toolkits/highwaystoolkit/6/bibliography/pdf/PBRMC-Latin_American_and_OECD_Experience_Zietlow.pdf.

## Chapter 12

AGC (American General Contractors). (2009). IPD: Lessons from the trenches. https://www.agc.org/news/2009/02/19/ipd-lessons-trenches.

Ahmad, I., Azhar, N., & Chowdhury, A. (2019). Enhancement of IPD characteristics as impelled by information and communication technology. *Journal of Management in Engineering*, 35(1), 04018055.

AIA (American Institute of Architects) California Council. (2007). Integrated project delivery: A working definition, Version 2, Sacramento, California.

AIA (American Institute of Architects). (2007). Integrated project delivery — a guide. California Council, AIA, Washington, DC.

AIA (American Institute of Architects). (2012). IPD case studies, Minneapolis.

AIA (American Institute of Architects). (2020). Integrated Project Delivery (IPD) family. https://www.aiacontracts.org/contract-doc-pages/27166-integrated-project-delivery-ipd-family.

Anvuur, A. and Kumaraswamy, M. (2007). Conceptual model of partnering and alliancing. *Journal of Construction Engineering and Management*, 133(3), 225–234.

Arcadis. (2019). Global construction disputes report, laying the foundation for success. https://www.arcadis.com/en/united-states/our-perspectives/2019/global-construction-disputes-report-2019/.

Assaad, R. and Abdul-Malak, M. A. (2020a). Timing of liquidated damages recovery and related liability issues. *Journal of Legal Affairs and Dispute Resolution in Engineering and Construction*, 12(2), 04520015.

Assaad, R. and Abdul-Malak, M. A. (2020b). Legal perspective on treatment of delay liquidated damages and penalty clauses by different jurisdictions: Comparative analysis. *Journal of Legal Affairs and Dispute Resolution in Engineering and Construction*, 12(2), 04520013.

Assaad, R. and El-Adaway, I. (2020b). A comprehensive management framework for preventing operational bankruptcy of construction firms. In *Construction Research Congress 2020: Project Management and Controls, Materials, and Contracts*. Reston, VA: American Society of Civil Engineers. pp. 49–57.

Assaad, R. and El-adaway, I. H. (2020a). Enhancing the knowledge of construction business failure: A social network analysis approach. *Journal of Construction Engineering and Management*, 146(6), 04020052.

Assaad, R. and El-adaway, I. H. (2021a). Determining critical combinations of safety fatality causes using spectral clustering and computational data mining algorithms. *Journal of Construction Engineering and Management*, 147(5), 04021035.

Assaad, R. and El-adaway, I. H. (2021b). Guidelines for responding to COVID-19 pandemic: Best practices, impacts, and future research directions. *Journal of Management in Engineering*, 37(3), 06021001.

Assaad, R. and El-adaway, I. H. (2021c). Impact of dynamic workforce and workplace variables on the productivity of the construction industry: New gross construction productivity indicator. *Journal of Management in Engineering*, 37(1), 04020092.

Assaad, R., Ahmed, M. O., El-adaway, I. H., Elsayegh, A., and Siddhardh Nadendla, V. S. (2021b). Comparing the impact of learning in bidding decision-making processes using algorithmic game theory. *Journal of Management in Engineering*, 37(1), 04020099.

Assaad, R., El-adaway, I. H., and Abotaleb, I. S. (2020a). Predicting project performance in the construction industry. *Journal of Construction Engineering and Management*, 146(5), 04020030.

Assaad, R., El-adaway, I. H., El Hakea, A. H., Parker, M. J., Henderson, T. I., Salvo, C. R., and Ahmed, M. O. (2020c). Contractual perspective for BIM utilization in US construction projects. *Journal of Construction Engineering and Management*, 146(12), 04020128.

Assaad, R., El-adaway, I. H., Hastak, M., and Needy, K. L. (2020d). Commercial and legal considerations of offsite construction projects and their hybrid transactions. *Journal of Construction Engineering and Management*, 146(12), 05020019.

Assaad, R., El-adaway, I. H., Baxmeyer, K., Harman, M., Job, L., and Lashley, H. (2021a). Allocation of risks and responsibilities in green and sustainable buildings. *Journal of Architectural Engineering*, 27(2), 04021002.

Assaad, R., Elsayegh, A., Ali, G., Abdul Nabi, M., and El-adaway, I. H. (2020b). Back-to-back relationship under standard subcontract agreements: Comparative study. *Journal of Legal Affairs and Dispute Resolution in Engineering and Construction*, 12(3), 04520020.

Association of Consultant Architects. (2008). Standard form of contract for project partnering, Kent, U.K.

Bilbo, D., Bigelow, B., Escamilla, E., and Lockwood, C. (2015). Comparison of construction manager at risk and integrated project delivery performance on healthcare projects: A comparative case study. *International Journal of Construction Education and Research*, 11(1), 40–53.

Cheng, E. and Li, H. (2002). Construction partnering process and associated critical success factors: Quantitative investigation. *Journal of Management in Engineering*, 18(4), 194–202.

Cheng, E. and Li, H. (2004). Development of a practical model of partnering for construction projects. *Journal of Construction Engineering and Management*, 130(6), 790–798.

Cheung, S. O., Yiu, K. T., and Chim, P. S. (2006). How relational are construction contracts? *Journal of Professional Issues in Engineering Education and Practice*, 132(1), 48–56.

ConsensusDocs (2020). 300 Series: Collaborative. https://www.consensusdocs.org/contract/300-2/.

Crane, T., Felder, J., Thompson, P., Thompson, M., and Sanders, S. (1999). Partnering measures. *Journal of Management in Engineering*, 15(2), 37–42.

Davison, J. and Ryan, C. (2008). Partnering contract review, Arup Project Management, Over Arup & Partners Ltd., London.

Donkor, S. (2011). Determinants of business failure: The perspective of SMEs building contractors in the Ghanaian Construction Industry. Ph.D. dissertation, School of Graduate Studies, Kwame Nkrumah University of Science and Technology.

El Asmar, M., Hanna, A. S., and Loh, W. Y. (2013). Quantifying performance for the integrated project delivery system as compared to established delivery systems. *Journal of Construction Engineering and Management*, 139(11), 04013012.

El-adaway, I. H. (2010). Integrated project delivery case study: Guidelines for drafting partnering contract. *Journal of Legal Affairs and Dispute Resolution in Engineering and Construction*, 2(4), 248–254.

El-adaway, I. H. (2013). Promoting the sustainability of relational contracting through addressing third party insurance obstacles. *Journal of Management in Engineering*, 29(3), 216–223.

El-adaway, I., Abotaleb, I., and Eteifa, S. (2017). Framework for multiparty relational contracting. *Journal of Legal Affairs and Dispute Resolution in Engineering and Construction*, 9(3), 04517018.

Ellison, D. and Miller, D. (1995). Beyond ADR: Working toward synergistic strategic partnership. *Journal of Management in Engineering*, 11(6), 44–54.

Elsayegh, A., El-adaway, I. H., Assaad, R., Ali, G., Abotaleb, I., Smith, C., Bootwala, M., and Eteifa, S. (2020). Contractual guidelines for management of infrastructure transportation projects. *Journal of Legal Affairs and Dispute Resolution in Engineering and Construction*, 12(3), 04520023.

Grajek, K., Gibson, E., and Tucker, R. L. (2000). Partnered project performance in Texas Department of Transportation. *Journal of Infrastructure Systems*, 6(2), 73–79.

Hanna, A. S. (2016). Benchmark performance metrics for integrated project delivery. *Journal of Construction Engineering and Management*, 142(9), 04016040.

Harmon, K. M. J. (2003). Dispute review boards and construction conflicts: Attitudes and opinions of construction industry members, Doctorate of Philosophy Thesis, Nova Southeastern University.

Ibrahim, M. W., Hanna, A., and Kievet, D. (2020). Quantitative comparison of project performance between project delivery systems. *Journal of Management in Engineering*, 36(6), 04020082.

Kent, D. C. and Becerik-Gerber, B. (2010). Understanding construction industry experience and attitudes toward integrated project delivery. *Journal of Construction Engineering and Management*, 136(8), 815–825.

Khalef, R., El-adaway, I. H., Assaad, R., and Kieta, N. (2021). Contract risk management: A comparative study of risk allocation in exculpatory clauses and their legal treatment. *Journal of Legal Affairs and Dispute Resolution in Engineering and Construction*, 13(1), 04520036.

Kumaraswamy, M. M., Ling, F. Y., Rahman, M. M., and Phng, S. T. (2005). Constructing relationally integrated teams. *Journal of Construction Engineering and Management*, 131(10), 1076–1086.

Lazar, F. (2000). Project partnering: Improving the likelihood of win/win outcomes. *Journal of Management in Engineering*, 16(2), 71–83.

Manata, B., Miller, V., Mollaoglu, S., and Garcia, A. J. (2018). Measuring key communication behaviors in integrated project delivery teams. *Journal of Management in Engineering*, 34(4), 06018001.

Miller, J. B., Garvin, M. J., Ibbs, C. W., and Mahoney, S. E. (2000). Toward a new paradigm: Simultaneous use of multiple project delivery methods. *Journal of Management in Engineering*, 163, 58–67.

Mosey, D. (2001). PPC 2000: The first standard form of contract for project partnering, Society of Construction Law, London.

National Institute of Building Sciences. (2015). Buildingsmart alliance releases NBIMS-US version 3. https://www.nationalbimstandard.org/buildingSMART-alliance-Releases-NBIMS-US-Version-3.

Neo, R. (1998). Partnering in project management. In *Proceedings of the Second International Conference on Construction Project Management: Critical Issues and Challenges into the Next Millennium*, pp. 19–20, Singapore: Centre for Advanced Construction Studies, School of Civi & Structural Engineering, Nanyang Technological University.

Noble, C. (2007). Can project alliancing agreements change the way we build. *Architectural Record*, 3(21), 5.

O'Connor, P. (2009). *Integrated Project Delivery: Collaboration Through New Contract Forms*. Minneapolis: Faegre & Benson, LLP.

Othman, A. A. E. (2008). Incorporating value and risk management concepts in developing low cost housing projects. *Emirates Journal for Engineering Research*, 13(1), 45–52.

Pheng, L. S. (1999). The extension of construction partnering for relationship marketing. Marketing Intelligence & Planning.

Pryke, S. D. (2006). Legal issues associated with emergent actor roles in innovative UK procurement: Prime contracting case study. *Journal of Professional Issues in Engineering Education and Practice*, 132(1), 67–76.

Saal, D., Villeneuve, D., Fisher, T. A., and Halevi, T. (2013). The faces of IPD/ID — a case study. *ASHRAE Transactions*, 119(2), 1–8.

Saunders, K. and Mosey, D. (2005). PPC 2000: Association of consultant Architects standard form of Project Partnering Contract. *Lean Construction Journal*, 2(1), 62–66.

SFAA (Surety and Fidelity Association of America). (2018). Why do contractors fail? https://www.surety.org/page/claimsresources.

Sive, T. (2009). Integrated project delivery: Reality and promise, a strategist's guide to understanding and marketing IPD. Alexandria, VA: Society for Marketing Professional Services Foundation.

Tserng, H. P., G. F. Lin, L. K. Tsai, and P. C. Chen. (2011). An enforced support vector machine model for construction contractor default prediction. *Automation in Construction*, 20(8), 1242–1249.

Zaghloul, R. and Hartman, F. (2003). Construction contracts: The cost of mistrust. *International Journal of Project Management*, 21(6), 419–424.

## Chapter 13

ASCE (2012). *Quality in the Constructed Project: A Guide for Owners, Designers, and Constructors*. Portland: American Society of Civil Engineers.

Baker, E., Mellors, B., Chalmers, S., and Lavers, A. (2009). *FIDIC Contracts: Law and Practice*. London: Routledge, Taylor & Francis Group.

Besaiso, H., Fenn, P., Emsley, M., and Wright, D. (2018). A comparison of the suitability of FIDIC and NEC conditions of contract in Palestine. *Engineering, Construction, and Architectural Management*, 25(2), 241–256.

Cao, Z. and Lumineau, F. (2015). Revisiting the interplay between contractual and relational governance: A qualitative and meta-analytic investigation. *Journal of Operations Management*, 33–34(1), 15–42.

Chen, C. (2008). Entry mode selection for international construction markets: The influence of host country related factors. *Construction Management and Economics*, 26(3), 303–314.

Chen, C. and Messner, J. I. (2009). Entry mode taxonomy for international construction markets. *Journal of Management in Engineering*, 25(1), 3–11.

Chen, C. and Messner, J. (2011a). Characterizing entry modes for international construction markets: Paving the way to a selection model. *Engineering Construction and Architectural Management*, 18(6), 547–567.

Chen, C. and Messner, J. I. (2011b). Permanent versus mobile entry decisions in international construction markets: Influence of home country- and firm-related factors. *Journal of Management in Engineering*, 27(1), 2–12.

Chen, C., Wang, Q., Martek, I., and Li., H. (2016a). International market selection model for large Chinese contractors. *Journal of Construction Engineering and Management*, 142(10), 04016044.

Chen, H., Li, X., Zeng, S., Ma, H., and Lin, H. (2016b). Does state capitalism matter in firm internationalization? Pace, rhythm, location choice, and product diversity. *Management Decision*, 54(6), 1320–1342.

Chen, Z., Li, H., Ren, H., Xu, Q., and Hong, J. (2011). A total environmental risk assessment model for international hub airports. *International Journal of Project Management*, 29(7), 856–866.

Chen, Y. Q., Wang, W. Q., Zhang, S. B., and You, J. Y. (2018). Understanding the multiple functions of construction contracts: The anatomy of FIDIC model contracts. *Construction Management and Economics*, 36(8), 472–485.

Cui, C., Liu, Y., Hope, A., and Wang, J. (2018). Review of studies on the public–private partnerships (PPP) for infrastructure projects. *International Journal of Project Management*, 36(5), 773–794.

Deng, X., Low, S. P., Li, Q., and Zhao, X. (2014). Developing competitive advantages in political risk management for international construction enterprises. *Journal of Construction Engineering and Management*, 140(9), 758–782.

Dikmen, I. and Birgonul, M. T. (2004). Neural network model to support international market entry decisions. *Journal of Construction Engineering and Management*, 130(1), 59–66.

Dikmen, I. and Birgonul, M. T. (2006). An analytic hierarchy process based model for risk and opportunity assessment of international construction projects. *Canadian Journal of Civil Engineering*, 33(1), 58–68.

Dikmen, I., Birgonul, M. T., and Han, S. (2007). Using fuzzy risk assessment to rate cost overrun risk in international construction projects. *International Journal of Project Management*, 25(5), 494–505.

Du, L., Tang, W., Liu, C., Wang, S., Wang, T., Shen, W., Huang, M., & Zhou, Y. (2016). Enhancing engineer–procure–construct project performance by partnering in international markets: Perspective from Chinese construction companies. *International Journal of Project Management*, 34(1), 30–43.

El-Sayegh, S. M. and Mansour, M. H. (2015). Risk assessment and allocation in highway construction projects in the UAE. *Journal of Management in Engineering*, 31(6), 04015004.

Folsom, R. H., Gordon, M. W., Spanogle, J. A., Fitzgerald, P. L., and Van Alstine, M. P. (2012). *International Business Transactions: A Problem-Oriented Coursebook* (11th ed.) Portland: Ringgold.

Gad, G. M. and Shane, J. S. (2017). Culture-risk-trust model for dispute-resolution method selection in international construction contracts. *Journal of Legal Affairs and Dispute Resolution in Engineering and Construction*, 9(4), 4517020.

Gao, R., Wu, J., Wang, H., and Lyu, S. (2020). International Market Selection of Chinese Construction Companies from the Perspective of Cultural Distance. In *ICCREM 2020: Intelligent Construction and Sustainable Buildings* (pp. 758–765). Reston, VA: American Society of Civil Engineers.

Geringer, J. M. and Hebert, L. (1989). Control and performance of international joint ventures. *Journal of International Business Studies*, 20(2), 23–254.

Girmscheid, G. and Brockmann, C. (2010). Inter- and intraorganizational trust in international construction joint ventures. *Journal of Construction Engineering and Management*, 136(3), 353–360.

Global Infrastructure Hub (2017). Global infrastructure outlook: Infrastructure investment needs 50 countries, 7sectors to 2040. Sydney, Australia: Oxford Economics.

Han, L. J., Zhang, S. B., Ma, P., and Gao, Y. (2019). Management control in international joint ventures in the infrastructure sector. *Journal of Management in Engineering*, 35(1), 04018051.

Han, S. H., Kim, D. Y., Kim, H., and Jang, W. S. (2008). A web-based integrated system for international project risk management. *Automation in Construction*, 17, 342–356.

Han, S. H., Diekmann, J. E., and Ock, J. H. (2005). Contractor's risk attitudes in the selection of international construction projects. *Journal of Construction Engineering and Management*, 131(3), 283–292.

Han, S. H., Kim, D. Y., and Kim., H. (2007). Predicting profit performance for selecting candidate international construction projects. *Journal of Construction Engineering and Management*, 133(6), 425–436.

Hastak, M. and Shaked, A. (2000). ICRAM-1: Model for international construction risk assessment. *Journal of Management in Engineering*, 16(1), 59–69.

Jia, R., Li, Q., Deng, X., Zhao, X., and Yuan, J. (2017). Entry mode taxonomy and choice of Chinese international construction companies. *Journal of Management in Engineering*, 33(3), 04016058.

Kapila, P. and Hendrickson, C. (2001). Exchange rate risk management in international construction ventures. *Journal of Management in Engineering*, 17(4), 186–191.

Klee, L. (2018). *International Construction Contract Law* (2nd ed). Hoboken: Wiely Blackwell.

Kim, D. Y., Ashuri, B., and Han, S. H. (2013). Financial valuation of investments in international construction markets: Real-options approach for market-entry decisions. *Journal of Management in Engineering*, 29(4), 355–368.

Lavikka, R., Kallio, J., Casey, T., and Airaksinen, M. (2018). Digital disruption of the AEC industry: Technology-oriented scenarios for possible future development paths. *Construction Management and Economics*, 36(11), 635–650.

Lee, K. W. and Han, S. H. (2017). Quantitative analysis for country classification in the construction industry. *Journal of Management in Engineering*, 33(4), 04017014.

Lee, R. M. (1988). International contracting-A formal language approach. In *Proceedings of the 21st Annual Hawaii International Conference on System Sciences*. Applications Track (Vol. 4, pp. 69-78). IEEE.

Li, G. H., Chen, C., Zhang, G., and Martek, I. (2020a). Bid/no-bid decision factors for Chinese international contractors in international construction projects. *Engineering Construction and Architectural Management*, 27(7), 1619–1643.

Li, G. H., Zhang, G., Chen, C., and Martek, I. (2020b). Empirical bid or no bid decision process in international construction projects: Structural equation modeling framework. *Journal of Construction Engineering and Management*, 146(6), 4020050.

Li, G. H., Chen, C., Martek, I., Ao, Y. B., & Dai, J. (2021). Bid or no-bid decision model for international construction projects: Evidential reasoning approach. *Journal of Construction Engineering and Management*, 147(2), 04020161.

Lim, H. W., Zhang, F., Fang, D., Pena-Mora, F., and Liao, P. C. (2021). Corporate social responsibility on disaster resilience issues by international contractors. *Journal of Management in Engineering*, 37(1), 04020089.

Lin, Y. H., Guo, Y., Kim, C. J., Chen, P. H., and Qian, M. (2020). The impact of relational governance on the adaptability of international contractors: A comparative study between China and Korea. *Engineering, Construction and Architectural Management*, 27(10), 3235–3259.

Liu, J., Zhao, X., and Yan, P. (2016). Risk paths in international construction projects: Case study from Chinese contractors. *Journal of Construction Engineering and Management*, 142(6), 05016002.

Liu, X. Y., Vredenburg, H., and Steel, P. (2014). A meta-analysis of factors leading to management control in international joint ventures. *Journal of International Management*, 20(2), 219–236.

Loo, S. C., Abdul-Rahman, H., and Wang, C. (2013). Managing external risks for international architectural, engineering, and construction (AEC) firms operating in gulf cooperation council (GCC) states. *Project Management Journal*, 44(5), 70–88.

Low, W. W., Abdul-Rahman, H., and Zakaria. N. (2015). The impact of organizational culture on international bidding decisions: Malaysia context. *International Journal of Project Management*, 33(4), 917–931.

Lu, W., Ye, M., Flanagan, R., and Ye, K. (2016). Corporate social responsibility disclosures in international construction business: Trends and prospects. *Journal of Construction Engineering and Management*, 142(1), 04015053.

Lumineau, F. and Henderson, J. E. (2012). The influence of relational experience and contractual governance on the negotiation strategy in buyer–supplier disputes. *Journal of Operations Management*, 30(5), 382–395.

McDermot, E., Agdas, D., Rodríguez Díaz, C. R., Rose, T., and Forcael, E. (2020). Improving performance of infrastructure projects in developing countries: An Ecuadorian case study. *International Journal of Construction Management*, 1–15. https://www.tandfonline.com/doi/abs/10.1080/15623599.2020.1797985?journalCode=tjcm20

Ndekugri, I., Smith, N., and Hughes, W. (2007). The engineer under FIDIC's conditions of contract for construction. *Construction Management and Economics*, 25(7), 791–799.

Neves, J. C. and Bugalho, A. (2008). Coordination and control in emerging international construction firms. *Construction Management and Economics*, 26(1), 3–13.

Preece, C. N., Isa, C. M. M., Saman, H. M., and Ibrahim, C. K. C. (2016). Development of entry location, entry timing and entry mode decision model for construction firms in international markets. *Construction Management and Economics*, 34(4–6), 1–22.

Shen, W., Tang, W., Wang, S., Duffield, C. F., Hui, F. K. P., and You, R. (2017). Enhancing trust-based interface management in international engineering-procurement-construction projects. *Journal of Construction Engineering and Management*, 143(9), 04017061.

Schepker, D. J., Oh, W., Martynov, A., and Poppo, L. (2014). The many futures of contracts. *Journal of Management*, 40(1), 193–225.

Steffey, R. W. and Anantatmula, V. S. (2011). International projects proposal analysis: Risk assessment using radial maps. *Project Management Journal*, 42(3), 62–74.

Tetteh, M. O. and Chan, A. P. (2019). Review of concepts and trends in international construction joint ventures research. *Journal of Construction Engineering and Management*, 145(10), 04019057.

Viswanathan, S. K. and Jha, K. N. (2019). Factors influencing international market selection for Indian construction firms. *Journal of Management in Engineering*, 35(5), 05019006.

Viswanathan, S. K. and Jha, K. N. (2020). Risk mitigation modelling of international construction projects executed by Indian firms: A structural equation modelling approach. *Engineering, Construction and Architectural Management*, 27(9), 2687–2713.

Viswanathan, S. K., Tripathi, K. K., and Jha, K. N. (2020). Influence of risk mitigation measures on international construction project success criteria–a survey of Indian experiences. *Construction Management and Economics*, 38(3), 207–222.

Viswanathan, S. K. (2020). Entry mode sequencing model for emerging international construction firms. *Journal of Management in Engineering*, 36(3), 05020002.

Wang, S. Q., Dulaimi, M. F., and Aguria, M. Y. (2004). Risk management framework for construction projects in developing countries. *Construction Management and Economics*, 22(3), 237–252.

Wang, T., Gao, S., and Liao, P. C. (2018). Systematic risk assessment and treatment framework of international construction project based on dynamic meta network analysis. *Construction Research Congress 2018: Construction Information Technology*, pp. 227–238.

Wang, T., Tang, W., Du, L., Duffield, C. F., and Wei, Y. (2016). Relationships among risk management, partnering, and contractor capability in international EPC project delivery. *Journal of Management in Engineering*, 32(6), 04016017.

Wang, C., Loo, S.C., Yap, J. B. H., and Abdul-Rahman, H. (2019). Novel capability-based risk assessment calculator for construction contractors venturing overseas. *Journal of Construction Engineering and Management*, 145(10), 04019059.

Williamson, O. E. (2002). The lens of contract: Private ordering. *American Economic Review*, 92(2), 438–443.

Wu, C. L., Fang, D. P., Liao, P. C., Xue, J. W., Li, Y., and Wang, T. (2015). Perception of corporate social responsibility: The case of Chinese international contractors. *Journal of Cleaner Production*, 107(16), 185–194.

Yarbrough, B. V. and Yarbrough, R. M. (1994). International contracting and territorial control: The boundary question. *Journal of Institutional and Theoretical Economics*, 150(1), 239–264.

Yu, T., Shen, G. Q., and Shi, Q. (2017). Comparing the performance quality of design-bid-build and design-build delivery methods. *Journal of Construction Engineering and Management*, 143(4), 04016111.

Zhao, Z. Y., Guo, X. L., and Chang, R. D. (2019). Market concentration and competitive intensity of the international engineering contracting industry. *Journal of Construction Engineering and Management*, 145(5), 04019026.

Zhao, Z. Y., Xu, K., Zuo, J., and Tang, C. (2017). Developing the international construction contracting market: Enterprise niche approach. *Journal of Management in Engineering*, 33(1), 04016027.

Zhang, S. B. (2018). Introductory remark. In Klee, L. (ed.), *International Construction Contract Law* (2nd ed.). Hoboken: Wiely Blackwell.

Zhang, S., Liu, X., Gao, Y., and Ma, P. (2019). Effect of level of owner-provided design on contractor's design quality in DB/EPC projects. *Journal of Construction Engineering and Management*, 145(1), 04018121.

Zhi, H. (1995). Risk management for overseas construction projects. *International Journal of Project Management*, 13(4), 231–237.

Zouher, A. E., Mohammed, A. A., Abdul-Rahman, H., and Kalsum, Z. U. (2014). Determining the relationship between conflict factors and performance of international construction projects. *Engineering, Construction, and Architectural Management*, 21(4), 369–382.

## Chapter 14

Adrian, J. J. (1988). *Construction Claims: A Quantitative Approach.* Champaign, IL: Stipes Publishing L.L.C.

Alaloul, W. S., Hasaniyah, M. W., and Tayeh, B. A. (2019). A comprehensive review of disputes prevention and resolution in construction projects. MATEC web of conferences,

Allen, R. K. (1998, July/August 1998). Professional practice survey results. *Journal of Management in Engineering*, 14(4), 30–34.

American Arbitration Association. (2003). Dispute-wise management. American Arbitration Association.

American Arbitration Association. (2004). AAA glossary of dispute resolution terms. Accessed November 1. http://www.adr.org/index2.1.jsp?JSPssid=15784.

Arcadis. (2020). Global construction disputes report: Collaborating to achieve project excellence. Accessed May 18. https://www.arcadis.com/en/knowledge-hub/perspectives/middle-east/2020/global-construction-disputes-report-2020.

Birr, J. and Scnchez, E. (2017). The most common commercial construction contract issues: Part I. Accessed May 20. https://www.jimersonfirm.com/blog/2017/03/common-commercial-construction-contract-issues-part-i/.

Braun, J. (1998, November 1998). Arbitration — promises made, promises broken. *Public Management*, 80(11), 9–11.

Brooker, P. and Lavers, A. (1997, November 1997). Perceptions of alternative dispute resolution as constraints upon its use in the UK construction industry. *Construction Management and Economics*, 15(6), 519–526.

Brown, H. J. and Marriott, A. L. (1993). *ADR: Principles and Practice*. London: Sweet & Maxwell.

Cakmak, P. I. (2021). The stakeholders' perspective on the factors contributing to construction disputes. *International Journal of Building Pathology Adaptation*, forthcoming. https://doi.org/10.1108/IJBPA-06-2020-0050.

Cheung, S.-O., Suen, H. C., and Lam, T.-I. (2002). Fundamentals of alternative dispute resolution processes in construction. *Journal of Construction Engineering and Management*, 128(5), 409–417.

Cheung, S. and Suen, H. (2002, October 2002). A multi-attribute utility model for dispute resolution strategy selection. *Construction Management and Economics*, 20(7), 557–568.

Cho, Y. (2020). Effective handling of construction disputes for strengthening the international competitiveness of the construction industry. *Korean Journal of Construction Engineering Management*, 21(4), 3–11.

Construction Industry Institute. (1995). Disputes Potential Index (SP23-3). The Construction Industry Institute.

Construction Industry Institute. (2017). SP23-3 — Disputes Potential Index, Version 2.0.

Diekmann, G. and Abdul-Hadi. (1994). DPI — Dispute Potential Index: A Study into the Predictability of Contract Disputes (SD-101). The Construction Industry Institute.

Diekmann, J. E. and Girard, M. J. (1995, December 1995). Are contract disputes predictable? *Journal of Construction Engineering and Management*, 121(4), 355–363.

Diekmann, J. E. and Nelson, M. C. (1985, March 1985). Construction claims: Frequency and severity. *Journal of Construction Engineering and Management*, 111(1), 74–81.

Dispute Resolution Board Foundation. (2005). DRBF Practices and Procedures. Accessed January 10. http://www.drb.org/.

Dispute Resolution Board Foundation. (2019). *Dispute Board Manual: A Guide to Best Practices and Procdures*. Charlotte, North Carolina: Dispute Resolultion Board Foundation.

Dispute Resolution Journal. (1997, Summer 1997). Big corporations often use ADR. *Dispute Resolution Journal*, 52(3), 7.

Editorials. (1985, May 30, 1985). Mediation picking up steam. Engineering News-Record, 23.

Editorials. (1994, July 11, 1994). Turning the tide on disputes has really started to pay off. *Engineering News-Record*, 233(28), 58.

Editorials. (1997, March 10, 1997). Measurement is the key. *Engineering News-Record*, 238(10), 62.

Editorials. (1999, February 15, 1999). Lightening up litigiousness. *Engineering News-Record*, 242(7), 68.

El-Sayegh, S., Ahmad, I., Aljanabi, M., Herzallah, R., Metry, S., and El-Ashwal, O. (2020). Construction disputes in the UAE: Causes and resolution methods. *Buildings*, 10(10), 171.

Federal Facilities Council & National Academy of Construction. (2004, September 23, 2004). Reducing construction costs: Uses of "Best Dispute Resolution Practices" by project owners. Washington, D.C.

Federal Facilities Council, & National Research Council. (2007). Reducing construction costs: Uses of best dispute resolution practices by project owners. Proceedings report. National Academies Press.

Fisher, R., Ury, W., and Patton, B. (1991). *Getting to Yes — Negotiation Agreement Without Giving* (2nd ed.). New York: Penguin Books.

Flanigan, K. (2000). *Insights in Construction: Fourth Report on the Nation's Leading Construction Companies*. New York: Deloitte & Touche LLP.

Flanigan, K., Darga, L., and Connors, M. (1997). Insights in construction: Third survey report on the nation's leading construction companies.

Frano, A. (1996, July/August 1996). Guarding against litigation. *Journal of Management in Engineering*, 12(4), 28–32.

Gebken, R. J. (2006). Quantification of transactional dispute resolution costs for the US construction industry.

Gebken, R. J. and Gibson, G. E. (2006). Quantification of costs for dispute resolution procedures in the construction industry. *Journal of Professional Issues in Engineering Education and Practice*, 132(3), 264–271. https://doi.org/10.1061/(ASCE)1052-3928(2006)132:3(264).

Goldberg, S. B., Sander, F. E. A., Rogers, N. H., and Cole, S. R. (1999). *Dispute Resolution: Negotiation, Mediation, and Other Processes* (3rd ed.). Los Angeles, California: Aspen Publishers.

Groton, J. P. (1997, Summer 1997). Alternative dispute resolution in the construction industry. *Dispute Resolution Journal*, 52(3), 48–57.

Groton, J. P. (2005, July 19–21, 2005). How to keep your project out of litigation, arbitration, and even mediation. In *CII Annual Conference — Leadership of Tomorrow: Bridging the Gap, Grapevine, TX.*

Harmon, K. M. J. (2003, October 2003). Resolution of construction disputes: A review of current methodologies. *Leadership and Management in Engineering*, 3(4), 187–201.

Hasanzadeh, S., Esmaeili, B., Gad, G. M., and Gransberg, D. D. (2018). Impact of owners' early decisions on project performance and dispute occurrence in public highway projects. *Journal of Legal Affairs and Dispute Resolution in Engineering and Construction*, 10(2), 04518004-04518001.

Hinchey, J. and Schor, L. (2002). The quest for the right questions in the construction industry. *Dispute Resolution Journal*, 57(3), 8–20. http://www.lib.utexas.edu:2048/login?url=http://search.epnet.com/login.aspx?direct=true&db=buh&an=7847161.

Ho, S. and Liu, L. Y. (2004, January/February 2004). Analytical model for analyzing construction claims and opportunistic bidding. *Journal of Construction Engineering and Management*, 130(1), 94–104.

Hughes, W., Champpion, R., and Murdoch, J. (2015). *Construction Contracts — Law and Management* (5th Ed.). New York: Routledge.

Javed, A. A., Pan, W., Chen, L., and Zhan, W. (2018). A systemic exploration of drivers for and constraints on construction productivity enhancement. *Built Environment Project and Asset Management*, 8(3), 239–252.

Jergeas, G. F. and Hartman, F. T. (1994, September 1994). Contractor' construction-claims avoidance. *Journal of Construction Engineering and Management*, 120(3), 553–560.

Keil, J. H. (1999, August 1999). "Hybrid ADR" in the construction industry. *Dispute Resolution Journal*, 54(3), 14–22.

Kellogg, J. C. (1999, Summer 1999). The contract disputes resolution continuum. Kellogg News, 1–5.

Kilian, J. (2003). *A Forensic Analysis of Construction Litigation, U.S. Naval Facilities Engineering Command University of Texas at Austin]*. Austin, TX.

Kilian, J. and Gibson, G. E. (2005, September 2005). Construction litigation for the U.S. Naval facilities engineering command 1982–2002. *Journal of Construction Engineering and Management*, 131(9), 945–952.

King, L. S., Kamarazaly, M. A. H., Hashim, N., Yaakob, A. M., and Man, N. H. (2019). Analysis of the issues of construction disputes and the ideal dispute resolution method. *Malaysian Construction Reseearch Journal*, 7(2), 153.

Komurlu, R. and Arditi, D. (2017). The role of general conditions relative to claims and disputes in building construction contracts. *New Arch-International Journal of Contemporary Architecture*, 4(2), 27–36.

Kovach, K. K. (2004). *Mediation: Principles and Practice* (3rd ed.). Eagan, MN: West Publishing Company.

Kumaraswamy, M. M. (1997). Conflicts, claims and disputes in construction. *Engineering Construction and Architectural Management*, 4(2), 95–111. http://www.blackwell-synergy.com/doi/abs/10.1046/j.1365-232X.1997.00087.x.

Kumaraswamy, M. M. (1998). Tracing the roots of construction claims and disputes. *COBRA '98 RICS Construction and Building Research Conference*, Oxford, UK.

Kurgan, T. J. (2005). *A Forensic Analysis of Construction Litigation, U.S. Army Corps of Engineers University of Texas at Austin]*. Austin, TX.

Lipsky, D. B. and Seeber, R. L. (1997). The use of ADR in U.S. corporations: Executive summary.

Loosemore, M. (1999, March 1999). Bargaining tactics in construction disputes. *Construction Management and Economics*, 17(2), 177–188.

Loosemore, M., Nguyen, B. T., and Denis, N. (2000, June 2000). An investigation into the merits of encouraging conflict in the construction industry. *Construction Management and Economics*, 18(4), 447–456.

Mastin, J. M., Nelson, E. L., and Robey, R. G. (2019). *Smith, Currie & Hancock's Common Sense Construction Law: A Practical Guide for the Construction Professional* (6th Ed.). New York: Wiley.

Matyas, R. M., Mathews, A. A., Smith, R. J., and Sperry, P. E. (1996). *Construction Dispute Review Board Manual*. New York: McGraw-Hill.

Mays, J. B. (2003). *Exploratory Review and Analysis of Dispute Resolution Techniques Employed in the Construction Industry by CURT Organizations University of Texas at Austin]*. Austin, TX.

McGeorge, D., London, K., Love, P., Davis, P., Jefferies, M., Ward, P., and Chesworth, B. (2007). Dispute avoidance and resolution a literature review. Cooperative Research Centre for Construction Innovation.

Michel, H. L. (1998, September/October 1998). The next 25 years: The future of the construction industry. *Journal of Management in Engineering*, 14(5), 26–31.

Mitropoulos, P. and Howell, G. (2001, May/June 2001). Model for understanding, preventing, and resolving project disputes. *Journal of Construction Engineering and Management*, 127(3), 223–231.

Mix, D. M. (1997). ADR in the construction industry: Continuing the development of a more efficient dispute resolution mechanism. 12 Ohio St. J. on Disp. Resol. 463.

Nelson, S. (2003). Advanced legal concepts — CE 395U.3 class notes. University of Texas at Austin.

Office of Government Commerce. (2002). Dispute resolution guidance. Office of Government Commerce.

Pappas, M. P. (2004, March 8). A flexible framework for the prevention and resolution of construction disputes. The Dispute Resolution Portfolio Program, Austin, TX.

Pawson, O. D. (2003, June/July 2003). Out of court — pros and cons of alternative dispute resolution tools. *Canadian Consulting Engineer*, 44(4), 53.

Peña-Mora, F., Sosa, C., and McCone, D. (2003). *Introduction to Construction Dispute Resolution*. Hoboken, New Jersey: Prentice Hall.

Rauzana, A. (2016). Causes of conflicts and disputes in construction projects. *Journal of Mechanical Civil Engineering*, 13, 44–48.

Reuben, R. C. (1996, August 1996). The lawyer turns peacemaker. *ABA Journal*, 82(8), 54–62.

Richter, I. and Mitchell, R. S. (1982). *Handbook of Construction Law and Claims*. Englewood Cliffs, New Jersey: Reston Publishing Company, Inc.

Richter, I. E. (2000). The project neutral: Neutralizing risk, maintaining relationships & watching the bottom line. *Construction Business Review*, 8(2), 52–54.

Rubin, R. A., Cohan, R. A., and Lurie, P. M. (1991). Beyond arbitration: Alternatives to binding dispute resolution in the construction industry. In K. M. Cushman (Ed.), *Construction Contracts and Litigation in the Economically Troubled 90's*, KM Cushman, ed. New York: Practicing Law Institute, pp. 959–985.

Rubin, R. A., Fairweather, V., and Guy, S. (1999). *Construction Claims: Prevention and Resolution* (3rd ed.). New York: John Wiley & Sons (1983).

Rubinstein, K. E. (2018). New ideas on how to cope. ENR: Engineering News-Record, 52–52. https://search.ebscohost.com/login.aspx?direct=true&AuthType=sso&db=a9h&AN=128005923&site=ehost-live&scope=site&custid=039-820.

Russell, J. S. and Jaselskis, E. J. (1992, December 1992). Predicting construction contractor failure prior to contract award. *Journal of Construction Engineering and Management*, 118(4), 791–811.

Semple, C., Hartman, F., and Jergeas, G. (1994, December 1994). Construction claims and disputes: Causes and cost/time overruns. *Journal of Construction Engineering and Management*, 120(4), 785–795.

Steen, R. H. and MacPherson, R. J. (2000, September/October 2000). Resolving construction disputes out of court. *Journal of Property Management*, 65(5), 58–60.

Stickel, A. I. (1999, August 1999). ADR will become growing factor in commercial disputes. *Corporate Legal Times*, 9(93), 38.

Stipanowich, T. J. (1987). Arbitration and the multiparty dispute: The search for workable solutions. 72 Iowa L. Rev. 473.

Stipanowich, T. J. (1995, November 1995). Dispute avoidance and resolution: The construction industry as the cutting edge of evolution. *The Construction Lawyer*, 15(4), 3–4.

Stipanowich, T. J. (1996a). Beyond arbitration: Innovation and evolution in the United States construction industry. 31 Wake Forest L. Rev. 65, 65–182.

Stipanowich, T. J. (1996b, April 1996). Charting the course: The 1994 construction industry survey on dispute avoidance and resolution — Part II. *The Construction Lawyer*, 16(2), 8–19.

Stipanowich, T. J. (2004, November 2004). ADR and the 'Vanishing Trial': The growth and impact of alternative dispute resolution. *Journal of Empirical Legal Research*, 1(3), 843–912.

Stipanowich, T. J. and O'Neal, L. K. (1995, November 1995). Charting the course: The 1994 construction industry survey on dispute avoidance and resolution — Part 1. *The Construction Lawyer*, 15(4), 5–12.

Treacy, T. B. (1995, January/February 1995). Use of alternative dispute resolution in the construction industry. *Journal of Management in Engineering*, 11(1), 58–63.

U. S. Census Bureau. (2020). Value of construction put in place at a glance. U.S. Census Bureau. Accessed December 9. https://www.census.gov/construction/c30/c30index.html.

Vorster, M. C. (1993). *Dispute Prevention and Resolution (SD-95)*. The Construction Industry Institute.

Wang, P. (2019). The Aetiology and Progression of Construction Disputes between Client and Contractor in the UK The University of Manchester (United Kingdom)].

Yates, D. J. (2003, June). Can claims and disputes (in construction contracts) be prevented or reduced? *Building Journal Hong Kong*.

Zack, J. G. (1997). Resolution of disputes: The next generation. *AACE International Transactions*, 50.

www.ingramcontent.com/pod-product-compliance
Lightning Source LLC
Chambersburg PA
CBHW081217180926
51613CB00004B/259
*9789811224775*